# TIMBER DESIGN

## for the
## Civil Professional Engineering Examination

### Second Edition

Robert L. Brungraber, P.E.

PROFESSIONAL PUBLICATIONS, INC.
Belmont, CA 94002

## In the ENGINEERING REVIEW MANUAL SERIES

Engineer-In-Training Review Manual
    Engineering Fundamentals Quick Reference Cards
    Mini-Exams for the E-I-T Exam
    E-I-T Index Card Of Figures and Tables
Civil Engineering Reference Manual
    Civil Engineering Quick Reference Cards
    Civil Engineering Sample Examination
    Civil Index Card of Figures and Tables
    Civil Engineering Review Course on Cassettes
    Seismic Design for the Civil P.E. Exam
    Timber Design for the Civil P.E. Exam
Structural Engineering Practice Problem Manual
Mechanical Engineering Review Manual
    Mechanical Engineering Quick Reference Cards
    Mechanical Engineering Sample Examination
    Mechanical Index Card of Figures and Tables
    Mechanical Engineering Review Course on Cassettes
Electrical Engineering Review Manual
Chemical Engineering Review Manual
    Chemical Engineering Practice Exam Set
Land Surveyor Reference Manual
Metallurgical Engineering Practice Problem Manual
Petroleum Engineering Practice Problem Manual
Expanded Interest Tables
Engineering Law, Design Liability, and Professional Ethics

## In the ENGINEERING CAREER ADVANCEMENT SERIES

How to Become a Professional Engineer
The Expert Witness Handbook—A Guide for Engineers
Getting Started as a Consulting Engineer
Intellectual Property Protection—A Guide for Engineers
E-I-T/P.E. Course Coordinator's Handbook

Distributed by: Professional Publications, Inc.
                1250 Fifth Avenue
                Department 77
                Belmont, CA 94002
                (415) 593-9119

**TIMBER DESIGN for the CIVIL P.E. EXAM**
**Second Edition**

Copyright © 1987 by Professional Publications, Inc. All rights are reserved. No part of this publication may be reproduced, stored in a retrieval system, or transmitted, in any form or by any means, electronic, mechanical, photocopying, recording, or otherwise, without the prior written permission of the publisher.

Printed in the United States of America

ISBN: 0-932276-73-3

Professional Publications, Inc.
1250 Fifth Avenue, Belmont, CA 94002

Current printing of this edition (last number) 6 5 4 3 2

# TABLE OF CONTENTS

Table of Contents .......................... v

Guide to Figures ........................... vii

Guide to Tables ............................ viii

Nomenclature .............................. ix

Subscripts ................................. ix

Superscripts ............................... x

Symbols ................................... x

Acknowledgements ......................... xi

1. Introduction ............................ 1

2. Wood Anatomy and Physical Properties . 3
Hardwoods and Softwoods ................. 3
Wood Anatomy ........................... 3
Lumber Defects ........................... 4
   Knots .................................. 4
   Cross Grain ............................ 5
   Checks, Shakes, Splits .................. 6
   Reaction Wood ......................... 7
Wood Density ............................ 7
   Amount of Wood ....................... 7
   Infiltrates ............................. 8
   Amount of Moisture .................... 8
   Shrinkage ............................. 9

3. Thermal Properties ..................... 10

4. Mechanical Properties of Wood ......... 12
Introduction .............................. 12
Strength and Stiffness: Test and Design Values ... 13
   Small Clear Specimen Method .......... 13
   Full-Size Testing ...................... 17

5. Commercially Available Lumber
   and Allowable Design Stresses .......... 18
Allowable Stress Tables .................... 19
   Species/Variety ........................ 19
   Moisture Content ...................... 19
   Grade and Size ........................ 19
   Sizes .................................. 19
Listed Properties .......................... 21
   Extreme Fiber Stress in Bending, $(F_b)$ ........ 21
   Tension Parallel to the Grain, $(F_t)$ ........... 21
   Horizontal Shear, $(F_v)$ ...................... 21
   Compression Perpendicular to the Grain, $(F_{c\perp})$ . 22
   Compression Parallel to the Grain, $(F_c)$ ....... 22
   Modulus of Elasticity, $(E)$ .................... 22
Glue-Laminated Allowable Stress Tables ......... 22

6. Connections ............................ 23
Introduction .............................. 23
Fastener/Grain Relationship
   and Load Application Direction ......... 23
   Combination of Load Types ............ 25
   Load Duration ........................ 26
   Species Grouping ...................... 26
   Condition of Use ...................... 26
Nails and Spikes .......................... 26
   Connection Capacity
      —Multiple Nails and Spikes ......... 28
   Lateral Loads ......................... 28
   Allowable Lateral Loads ............... 30
   Withdrawal Loads .................... 30
Bolts, Lag Screws, Shear Plates, and Split Rings . 36
   Capacity Reduction with Multiple Fasteners .... 36
   Condition of Use ...................... 38
   Net Section ........................... 39
   Angle of Load to Grain ................ 39
Lag Screws ............................... 40
   Withdrawal Loads .................... 40
   Lateral Loads in Lag Screws ........... 40
Bolts ..................................... 45
   Reading Load Tables .................. 46
   Load Distribution as Influenced by
      Connection Details .................. 50
Spacing Requirements for Bolts and Lag Screws .. 52

Minimum Full-Strength Spacing ............... 52
Shear Plates and Split Rings (Connectors) ....... 61
    Considerations in Connector Design ........... 62
    Edge Distance ............................... 65
    End Distance ................................ 65
    End Distance Without Square-Cut Ends ....... 65
    Connector Spacing ........................... 66

## 7. Review of Essential Mechanics of Materials Concepts ........ 86
Introduction ..................................... 86
Axial Loadings ................................... 86
    Buckling .................................... 87
    Axial Deflections ............................ 89
Transverse Loading ............................... 90
    Bending Stresses ............................ 90
    Shear Stresses .............................. 90
    Deflections in Transversely-Loaded Members ... 91
    Lateral Stability of Bending Members ......... 92
Combined Loadings ............................... 92
    Transverse and Axial Tension Loading ......... 93
    Transverse and Axial Compression Loading ..... 93
    Eccentric Loading ........................... 94

## 8. Design of Timber Structural Members —Axial Members ...................... 94
Tension Members ................................. 94
Compression Members ............................. 95
    Connection at a Braced Point ................. 96
    Connection at an Unbraced Point ............. 96
    $F_c'$, Allowable Compression Stress, Considering Stability ........................ 96
    Column Slenderness Categories ................ 97
    Spaced Columns ............................. 103
    Built-up Columns ........................... 104

## 9. Design of Timber Structural Members —Bending Members .................... 105
Bending Stresses ................................. 105
    Allowable Bending Stresses, Lateral Buckling .. 105
    Beam Slenderness Categories ................. 107
    Allowable Bending Stresses, Limited by Extreme Fiber Failure ...................... 107
    Curved Bending Members .................... 108
Shear Stresses in Bending Members ............... 109
    Shear With Loads Near Supports ............. 110
    Shear in Notched Beams .................... 110
    Shear in Connections ....................... 111
    Shear in Checked Beams .................... 111

Beam Deflections ................................ 112
Bearing Stresses at Loads and Supports ......... 112

## 10. Plywood ................................ 119
Introduction ..................................... 119
    Plywood Grades and Types ................... 119
    Plywood Structural Applications .............. 120
Plywood Section Properties ....................... 121
    Direction of Face Grain ...................... 121
    Thickness for All Properties Except Shear ..... 121
    Thickness for Shear ......................... 121
    Cross-Sectional Area ........................ 124
    Moment of Inertia .......................... 124
    Section of Modulus ......................... 124
    Rolling Shear Constant ...................... 125
Allowable Stresses–Plywood ...................... 126
    Grade Stress Levels ......................... 126
    Plywood Species Groups .................... 126
    Conditions of Use .......................... 131
    Allowable Stress Table ...................... 131
    Duration of Load Factors .................... 131
Diaphragms and Shear Walls ..................... 131
    Diaphragms ................................ 132
    Shear Walls ................................ 133
    Design Method–Shear Walls and Diaphragms . 134
Plywood-Lumber Built-Up Beams ................ 139
    Design Consideration ....................... 140
    Trial Section ............................... 142
    Lumber Flanges ............................ 142
    Plywood Webs ............................. 143
    Flange to Web Connection .................. 143
    Deflections ................................ 143
    Details .................................... 144
    Lateral Stability ............................ 144
    Section Properties .......................... 146

## 11. Sample Problems ....................... 151

## 12. Appendices ............................. 183

## 13. References ............................. 197

## 14. Index .................................... 199

# GUIDE TO FIGURES

| | | |
|---|---|---|
| F1 | Knot Shapes | 5 |
| F2 | Checks, Shakes, and Splits | 6 |
| F3 | Characteristic Shrinkage and Distortion | 10 |
| F4 | Orientation Axes | 12 |
| F5 | Relation of Strength to Duration of Load | 16 |
| F6 | Fastener Installation | 24 |
| F7 | Load Application | 24 |
| F8 | Load Application Relative to Grain Direction | 25 |
| F9 | Load-Deformation Curve for Typical Laterally Loaded Nail | 29 |
| F10 | Number of Fasteners in a Row | 37 |
| F11 | Seasoning in Place | 39 |
| F12-15 | Loaded Truss | 50 |
| F16 | Spacing Dimensions–Bolt and Lag Screws | 52 |
| F17 | Types of Split Rings and Shear Plates | 62 |
| F18 | Number of Faces with Connectors | 64 |
| F19 | End Distance in Members Without Square Ends | 65 |
| F20 | Connector Spacing and Angle to the Grain | 66 |
| F21-30 | Load Data and Charts for Connectors | 71 |
| F31 | Member with Axial Load | 87 |
| F32 | Column Subject to Buckling | 88 |
| F33 | Column Failure Stresses | 89 |
| F34 | Transversely-Loaded Member | 90 |
| F35 | Longitudinal Bending Stresses | 90 |
| F36 | Shear Stresses Due to Transverse Loads | 91 |
| F37 | Member Subjected to Combined Loading | 92 |
| F38 | Column with Eccentric Load | 94 |
| F39 | Slenderness Ratio for a Column | 97 |
| F40 | Effective Beam Lengths | 106 |
| F41 | Loads Near Supports | 108 |
| F42 | Shear Stresses in Notched Beams | 111 |
| F43 | APA Grade-Trademark Stamp | 120 |
| F44 | Plywood in Bending | 125 |
| F45 | Shear Stress Orientations | 125 |
| F46 | Species Group–Span Rating Relationship | 129 |
| F47 | Plywood-Lumber Beam Cross-Sections | 140 |

# GUIDE TO TABLES

| | | |
|---|---|---|
| T1 | Thermal Conductivities | 11 |
| T2 | Common Load Duration Factors | 16 |
| T3 | Species Groupings for Connection Design | 27 |
| T4 | Condition of Use Factors (CUF) for Connections | 28 |
| T5 | Common Nail and Spike Dimensions | 28 |
| T6 | Design Values–Nails and Spikes in Lateral Loads | 31 |
| T7 | Allowable Withdrawal Loads–Nails and Spikes | 32 |
| T8A | Modification Factor, K, for Wood Side Plates | 37 |
| T8B | Modification Factor, K, for Metal Side Plates | 38 |
| T9 | Special Condition of Use Factors for Connections Seasoning in Place | 39 |
| T10 | Allowable Withdrawal Loads–Lag Screws | 41 |
| T11 | Allowable Lateral Loads in Lag Screws–Wood Side Pieces | 42 |
| T12 | Allowable Lateral Loads in Lag Screws–Metal Side Pieces | 43 |
| T13 | Lateral Loads in Bolts–Design Values | 48 |
| T14 | Capacity Modification Using Lag Screws in Connectors | 63 |
| T15 | Species Groupings for Shear Plates and Split Rings | 67 |
| T16 | Effective Column Lengths | 89 |
| T17 | Percent Reductions in Strength for Built-Up Columns | 104 |
| T18 | Size Factors for Glue-Laminated Beams | 108 |
| T19 | Allowable Radial Compression–Curved Glue-Laminated Beams | 110 |
| T20 | Effective Plywood Section Properties | 122 |
| T21 | Plywood Specifications–Abbreviated | 127 |
| T22 | Plywood Species Classifications | 129 |
| T23 | Plywood Allowable Stresses | 130 |
| T24 | Required Panel Details–Diaphragms | 132 |
| T25 | Plywood Shear Wall Capacities | 133 |
| T26 | Preliminary Capacities of Plywood-Lumber Beam Cross-Sections | 141 |
| T27 | Area of Unspliced Flange Members | 143 |
| T28 | Bending Deflection Increase to Account for Shear | 144 |
| T29 | Lateral Bracing Required for Plywood-Lumber Beams | 144 |

# TIMBER DESIGN

## NOMENCLATURE

A    member cross sectional area ($in^2$)
b    width of cross section (in)
C    modification factor, specific to subscript
c    distance to extreme fiber of section (in)
CUF    condition of use factor
d    depth of cross section or diameter (in)
E    Young's modulus of elasticity (psi)
e    eccentricity (in)
F    tabulated allowable stress (psi)
f    actual stress at a cross section (psi)
G    specific gravity
I    cross sectional moment of inertia ($in^4$)
J    factor which includes the slenderness ratio
K    slenderness ratio separating long and intermediate column behavior
L    member length (in)
LDF    load duration factor
M    bending moment at the section (in-lb)
MC    moisture content (%)
N    load on fastener(s) at angle to grain between $0°$ and $90°$
P    load (psi)
Q    first moment of cross sectional area beyond y, about the neutral axis ($in^3$)
R    radius of curvature (in) or (ft)
S    section modulus = $I/c = bd^2/6$ ($in^3$)
V    shear force at cross section (lb)
WGT    weight of a sample (lb)
y    distance from neutral axis to point where shear stress is being evaluated (in)
Y    deflection (in)

## SUBSCRIPTS

1    in fastener group analysis, area of main member(s), before drilling
2    in fastener group analysis, area of side member(s), before drilling
b    bending stress (or length of bearing, when used with L)

# Timber design

| | |
|---|---|
| c | compression parallel to the grain |
| c⊥ | compression perpendicular to the grain |
| c | curvature factor, when used with C |
| DRY | oven-dry condition |
| e | effective, as in column length |
| F | size factor, when used with C |
| f | form factor, when used with C |
| g | end grain bearing stress |
| k | dividing value between long and intermediate beams, when used with C |
| n | specific angle with grain for evaluated property, or net; as in net moment of inertia |
| r | radial stress |
| rc | radial compression |
| rt | radial tension |
| t | tension along the grain, or total; as in total moment of inertia |
| v | shear stress |
| WET | moisture content being evaluated |

## SUPERSCRIPTS

'    allowable stress after adjustments for slenderness

## SYMBOLS

$\theta$    angle between the longitudinal wood axis and the load or stress direction

## Acknowledgements

I would like to thank my parents for providing me with support, both financial and spiritual, throughout my academic career. Colorado State University's fine Timber Structures Program gave me the specific background required for this work. The Construction Management Program at Stanford was extremely generous with their computer and printer facilities. The many revisions in the second edition version of this work (occasioned by author errors and changes in the timber code) were accomplished only through the tremendous efforts of Major William W. Watkin III, an intelligent and dogged editor and student, as well as a valued friend. Finally, profound thanks go to my wife Joel, for her unflagging encouragement and willingness to accomodate this work in our lives.

RESERVED FOR FUTURE USE

# 1. INTRODUCTION

This book is primarily intended to help you prepare for timber structural questions on the professional engineering exam. While this book was written to be self-contained, the National Design Specifications published by the National Forest Products Association is the basic timber code used in the United States and its use is highly recommended.[1]

For those engineers who design timber structures in practice, this book will be a useful reference due to its worked examples illustrating the principles found in the code. The book is necessarily problem oriented and, therefore, cannot be exhaustive in its treatment of wood science. If, for example, you are interested in sophisticated wood mechanics, there are several references listed in the bibliography.

The interesting thing about the need for this book is that questions on timber structures have reappeared on P.E. exams after a lengthy hiatus. Why the renewed emphasis on the material after so many years? The answer is that timber has moved beyond the realm of residential and temporary construction areas that neither required nor received much engineering attention.

Many structures being built in the U.S. with wood today would have been made of steel or concrete only a few years ago. The major

---

[1] National Forest Products Association, 1619 Massachusetts Avenue, N.W., Washington, D.C. 20036

advance has been in glue-laminating large members from smaller, standard size lumber. With its high strength-to-weight ratio, this glue-laminated timber now dominates the long-span rigid dome market.

The proven durability of preservative-treated timber has opened up a large replacement bridge market in short to medium span ranges. Because of economic and schedule considerations, more and more low-rise commercial buildings are being framed with timber joists and rafters.

Timber's unique characteristics have made it the material chosen for immense and sophisticated structures, such as the 200 foot wide, 590 foot long Air Force trestle in New Mexico which holds a B-52 124 feet in the air for testing. Finally, recent increases in building rehabilitation have involved engineers in the evaluation and strengthening of older timber structures.

Beyond its primary role as the world's most widely used fuel, timber is also the most common building material -- both in number of structures and total floor area. It is easy to believe that it was the first material used for above-ground shelters. The surprise is that timber design methods are as relatively simplistic as they are after all these years.

Timber's characteristics and technolgy's timing have combined to hold back the development of timber analysis from the sophistication levels achieved in steel and concrete. Timber is the only organic building material. Its natural source not only makes it renewable, but in many ways also the most complex of building materials. Even the most powerful computer analysis programs can be overwhelmed by the complete response of a timber structure. Timber has also been overlooked by engineers because, as they were finding ways to deal with complex problems, timber was being replaced in large structures by steel and concrete -- simpler materials with which to deal.

The timber codes are changing rapidly, however, as new products are introduced, larger and longer timber members are used, and more powerful analytical tools are developed. So, while the present codes may not address issues that arise frustratingly often, engineers should appreciate the codes for the freedom available, supplementing these codes with current research when required.

## 2. WOOD ANATOMY AND PHYSICAL PROPERTIES

Wood is an extremely complex organic material. Scientists have been studying its structure for centuries. Consequently, the fields of wood anatomy and physics of behavior are immense and growing. In order to keep this book focused on structural applications, the wood anatomy presented will be very superficial, certainly below the detail level of high school biology. Only those features and orientations of wood's internal structure which significantly effect its behavior as a building material will be covered.

### A. HARDWOODS AND SOFTWOODS

"Hardwood" and "softwood", while being commonly used and understood terms, are often misnomers. Hardwoods such as balsa and aspen are quite soft, while softwoods such as white pine and true fir are quite hard. The less common, but more precise terms, "deciduous" and "coniferous", are not universally applicable either. Deciduous (hardwood) trees are defined as those which lose their broad leaves annually, a rule violated regularly in the tropics. Coniferous (softwood) trees, on the other hand, are not supposed to lose their leaves (needles) and should have seeds in cones. Again, there are species commonly recognized as softwoods which ignore these definitions. But, the hardwood/softwood distinction is so widespread that time spent disputing it is largely wasted.

There are other differences between the two tree types. The evolution of softwoods is longer, and their structure is far simpler than that of the hardwoods. There are only 650 species of softwoods, and they can be found in vast tracts containing only one species. There are over 250,000 species of hardwoods, however, and they are widely scattered and interspersed.

The entire issue of softwood versus hardwood is largely an academic one, however, as far as the structural applications of this book are concerned. The vast majority of structural timber requirements are filled with softwoods. Hardwood is used for fuel, railroad ties, pallets, furniture and paneling, and many other non-structural products. Exceptions to this pattern are the older structures built amid hardwood forests of the Eastern United States.

### B. Wood Anatomy

The single crucial thing to grasp about wood anatomy, as it affects the behavior and design of timber members and connections, is that wood is basically a bundle of tubes bound together by a natural glue. The tubes are wood cells aligned along the major axes of the tree and its branches. These cells are an eighth to a half inch long with walls composed largely of cellulose. The "glue" binding the cells together is lignin, a very complex and tough compound.

As a tree grows, its cells are laid up in concentric layers under the bark. The bark splits and expands, making room for the new wood. Once a cell is in place, it does not move within the tree. Fences nailed to trees are encased by them, but not lifted. Except in the tropics, the types of new cells are not the same throughout the year. The differences in spring and summer wood give rise to wood's growth rings and grain.

### C. Lumber Defects

A tree grows and responds to its environment in many ways. It inevitably develops certain characteristics that reduce the strength of the lumber cut from it. The worst of these defects are knots, cross grain, checks and splits, and reaction wood.

#### Knots

Knots are the most obvious and critical lumber flaw. A knot is a section of branch that has been enveloped by the trunk of the tree as it expanded. If the branch was an early and low one that had died and been encased along with its dead bark, the resulting knot is said to be "loose" or "encased." A branch that was alive when the tree was cut leaves "intergrown" or "tight" knots in the lumber, with uninterrupted contact between wood cells.

The shape of a knot in lumber is a function of how the plane of the cut intersected the branch. A cut normal to the axis of the branch results in a "round knot." If the cut is nearly parallel to the branch axis, the knot is called a "spike knot." Any orientation between normal and parallel makes "oval knots."

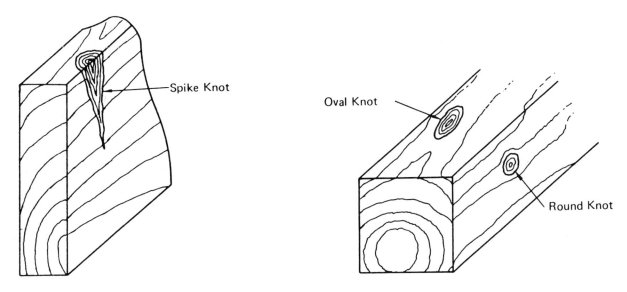

**Figure 1
Knot Shapes**

Knots reduce lumber's strength in several ways. There is an interruption of the fibers that would otherwise be aligned along the strong axis of the piece. Even the fibers that are left deviate from that axis in order to go around the knot. Finally, there are significant stress concentrations in the vicinity of the knot.

### Cross Grain

Cross grain is a general term describing wood fibers that are not aligned with the member's major axis. Cross grain is caused by cutting trees that have grown spirally or with severe taper. Some cross grain is the inevitable result of passing parallel planes through a tapered body. Cross grain drastically reduces the tensile strength of a member and can precipitate unexpected and early bending failures.

The most common measure of cross grain is the slope of grain. Slope of grain is the angle, expressed as a ratio, between the wood fibers and the longitudinal axis of the lumber. Large slope of grain is second only to knots in reducing the grade of a piece of lumber.

### Checks, Shakes, and Splits

Checks and splits are fractures in the wood that open as the lumber dries. Checks align with the longitudinal axis of the piece and are normal to the growth rings. Splits are generally wider cracks along the axis and are usually found at the member ends. The worst of the splits are often trimmed off after the lumber dries as the final step in cutting lumber.

Shakes are thin voids within the tree which seem to occur naturally as the tree is growing. Shakes also align with the axis of the tree, but they are in plane with the growth rings. It is as if the tree had delaminated slightly at a transition between spring and summer wood.

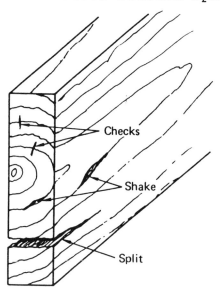

**Figure 2**
**Checks, Shakes, and Splits**

Timber's strength relies upon the connection between adjacent wood cells. Though checks, shakes, and splits may close, the cells never reconnect across the fracture. By functionally separating a piece of lumber, checks, shakes, and splits are detrimental in many ways. Just what and how much effect these separations have depends on the location and severity of the defects, and the use of the timber.

Reaction Wood

Reaction wood is created as a response by the tree to its changing dead load. As branches get longer and heavier, or as the tree is tilted for some reason, the wood is subjected to bending stresses. The tree responds by growing thicker-walled, denser cells at these locations. Most softwoods grow "compression wood" on the underside of the branches. Most hardwoods, however, grow "tension wood" on branch topsides. The denser reaction woods shrink more as they dry than the surrounding normal wood, resulting in dramatic and abrupt warps in any lumber that contains areas of reaction wood.

## D. Wood Density

The density of wood varies a great deal, even within a single species. Increased density is the most reliable indicator of a species' higher strength. Density is primarily a function of the amount of wood present, infiltrates, and the amount of moisture present.

Amount of Wood

The specific gravity of the basic wood material, the cell walls, is a nearly constant 1.53, regardless of species. What does vary is the amount of that material present in the tree. Oak cells have relatively thick walls, while balsa cell walls are very thin.

Within a species, growth rate is the major variable in establishing density. The slower a softwood grows, the higher its percentage of the denser summer wood and subsequent overall density. A tree's growth rate is affected by local environmental conditions more than anything else.

### Infiltrates

Only cells within several inches of the bark are kept alive to transfer fluids within the tree. As the tree grows, this "sapwood" gradually transforms into "heartwood." A part of that transformation process involves the deposition on the cell walls of chemicals known as infiltrates or extractives. There is wide variety in types and amounts of infiltrates. Silicas in tropical woods, for example, dull saw blades rapidly. The most rot-resistant timbers result from the presence of certain infiltrates. The densest tropical woods, some of which will sink in water, have large amounts of infiltrates.

### Amount of Moisture

The amount of water present is the major influence on the density of a given piece of wood. Cellulose is a hygroscopic material, meaning it absorbs and gives up water readily. Since wood is primarily made of cellulose, it changes moisture content in response to changing conditions.

The most common measure of the amount of water present in wood at any time is moisture content. Moisture content varies widely with species and time. Fresh cut cedar can have a moisture content as high as 250%.

$$MC = \frac{(100)(WGT_{wet} - WGT_{dry})}{WGT_{dry}} \qquad (1)$$

The water in wood is in two forms. The majority of the water is known as "free water" and simply fills the hollow cells while the tree is alive. This is the first water to leave after the tree is cut, and only the density of the wood changes as it goes. There is about five times as much free water as there is "bound water." The bound water is chemically bonded in the walls of the wood cells. As the bound water leaves, the wood shrinks and generally gets stronger.

There are some specific moisture content levels that are of interest.

- Green: When the tree is first cut, its moisture content starts to drop as the free water is lost. The term "green" wood can mean the fresh-cut state. Green wood is also defined in the allowable stress tables as having a moisture content level of 19% and above.

- Fiber Saturation Point (FSP): Once all the free water is gone, but none of the bound water is yet lost, the moisture content of the wood is defined as being at the "fiber saturation point." The fiber saturation point of all species is about 30%.

- Equilibrium Moisture Content (EMC): The wood continues to dry out by losing bound water until it achieves an "equilibrium moisture content." The equilibrium moisture content of wood varies with species, ambient temperature, and humidity. In the United States, equilibrium moisture content can range from 5% to 25%, with 12% to 15% being the common range.

- Oven-Dry: In order to lower the moisture content below its equilibrium level, the water must be driven off. The wood is heated in a ventilated oven at a temperature just above water's boiling point, and the weight is monitored. There is no absolutely dry wood, but once a specimen has reached a relatively stable weight, it is said to be "oven-dry."

Shrinkage

The amount wood shrinks as it goes from the fiber saturation point to oven-dry is also a function of direction relative to the fibers. Luckily for timber structures, the shrinkage along the fibers (longitudinally) is very small -- 0.1 to 0.2%. This means that a ten foot long timber will shorten less than an eighth of an inch.

Since it is the cell wall that shrinks with decreasing moisture content, it is not surprising that wood shrinks much more across the cells than along them. There is a difference in the shrinkage across the grain, which depends on the orientation of the growth rings. Along the rings (the tangential direction), wood shrinks about 7% from fiber saturation point to oven-dry. Across the rings (the radial direction), wood shrinks 3 or 4%. An eight inch wide timber might, therefore, shrink as much as a quarter inch from fiber saturation point to equilibrium moisture content. This dimension change with change in local conditions can cause problems for unsophisticated designers. Connection detailing is the area with the most potential for problems.

The difference in shrinkage between the two cross-grain directions can also cause some difficulties. The differential shrinkage means that the circumference shrinks less than the radius. The resultant hoop tensions in the wood make longitudinal splits in dry timber inevitable. Since lumber is typically cut from green wood and then dried, the differential shrinkage also causes some distortion of the cross sections.

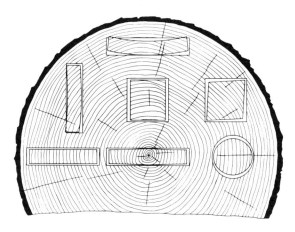

Figure 3
Characteristic Shrinkage and Distortion

Since both volume and weight are functions of the amount of water present in wood, comparing species on the basis of density can be misleading and confusing. Specific gravity is the parameter commonly used for comparing species with each other. Specific gravity is the ratio of the wood's density to that of water. The wood density must be derived from weights and volumes evaluated at some specified moisture content. Unfortunately, several benchmark conditions have been used over the years. Since the structural codes generally use oven-dry for both weight and volume, that is the condition assumed in this book.

## 3. THERMAL PROPERTIES

The growing interest in energy efficiency has focused attention on the thermal performance of all building materials, and wood is no exception. Most of the recent research is well beyond the structural scope of this book. A material's response to dynamic environmental conditions is a complex function of several material properties. Table 1 shows that the thermal conductivity of wood is far less than that of any other structural material.

**Table 1**
**Representative Thermal Conductivities**

| Material | Thermal Conductivity, k $Btu/hr\text{-}ft^2\text{-}in\text{-}{}^\circ F$ |
|---|---|
| Air | 0.168 |
| Wood (softwood average) | 0.80 |
| Glass | 5.5 |
| Concrete | 12.6 |
| Steel | 312.0 |
| Aluminum | 1,416.0 |

The dimensional changes wood undergoes with variation in moisture content are far greater than those caused by temperature changes. Elevated temperatures drive off water and shrink wood, while the thermal response is to expand. The accepted practice, in face of these conflicting effects, is to neglect thermal expansion and contraction.

One characteristic of timber's thermal properties which should interest structural engineers is the larger members' resistance to fire. Even though wood supports combustion at lower temperatures than other structural materials, it still needs oxygen and elevated temperature to do so. Wood is such a good insulator that a large timber can be in an inferno while the temperature is elevated only slightly inside the member itself. Consequently, timbers char and lose cross-section and capacity very slowly. The lengthened and increased endurance of wood in fires often results in reduced fire insurance rates for heavy timber structures.

# 4. MECHANICAL PROPERTIES OF WOOD

## A. Introduction

As described earlier, wood consists of cells laid up in concentric layers around the tree. As a result of this and its growth irregularities, wood is an anisotropic material. This means that all of wood's mechanical properties will be functions of position and orientation.

A commonly used simplification is an assumed set of three perpendicular axes of symmetry. Wood can then be treated as an orthotropic material, with only three directions with which to deal. The axes are related to the growth rings and were mentioned previously when describing differential shrinkage. The three axes are commonly designated as longitudinal, radial, and tangential.

The longitudinal axis is parallel to the wood fibers, and is sometimes called the "strong axis." The radial axis has its origin at the center of the tree and points outward. In a rectangular piece of lumber, the radial axis is in the cross-section and is perpendicular to the growth rings. The last axis is in the tangential direction, following the growth rings. In a tree, the tangential axis is circumferential; but, in a piece of lumber, it is assumed to lie in a plane parallel to the rings.

By cutting trees in different ways, sawmills can produce planks with the tangential and radial axes in a variety of positions. The two basic options, "plain-sawed" and "quarter-sawed", are shown in figure 4.

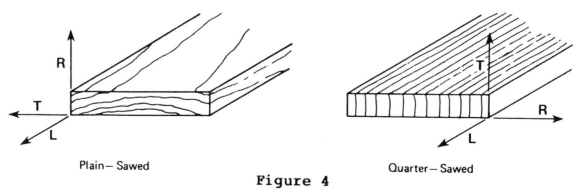

**Figure 4**
**Orientation Axes**

Though the orthotropic model reduces the anisotropic wood to one with only three directions, designing with wood still remains more complex than with other common structural materials. For example, there are six separate Poisson's ratios combining the three different directions for load application and strain measurement. There are also three possible compressive stress directions.

In order to simplify these orientation options to a manageable level, some assumptions and groupings are widely used. The radial and tangential directions are usually lumped into a "cross-grain" category. Some properties, such as modulus of elasticity, are only needed for the longitudinal direction. Finally, certain properties, such as the Poisson's ratio, are so unlikely to concern structural designers that they are usually neglected.

### B. Strength and Stiffness: Test and Design Values

Timber design has traditionally involved comparing design stresses and loads with those judged to be allowable. Limit-state design procedures are being developed, but they have yet to appear in the codes. Timber strength depends on many different parameters. Strength varies with species, within species, within species from the same stand, and even within the same tree. Clearly, establishing safe design values for such a variable material is a paramount issue.

There are two basic methods used to establish working stresses for timber design. The older method, historically used in the United States, is to test large numbers of small clear specimens in the lab and correct for the defects and service conditions that are found in actual use. The newer method, developed in Europe and gaining acceptance here, is to test the full-size timbers in nondestructive ways and to correct only for actual service conditions -- not for defects.

#### Small Clear Specimen Method

This method relies completely on the validity of the test specimen sampling methods, which are correspondingly precise and complicated. A large enough number of tests are run to give meaningful values.

Even with "identical looking" small clear specimens there is a spread in the test results. The corrections for defects and service are made to a strength level below which only five percent of the tested specimens failed. This starting point for corrections is known as the "five percent exclusion limit."

There is a variety of correction factors applied to this 5% percent exclusion limit. In summary, they are:

- Defects: The American Society of Testing and Materials includes in its Standard D245 an exhaustive list of "strength ratios" to be applied to the 5% exclusion limit on the basis of various defects. Strength ratios vary from 4 to 99%, depending on the relative severity of the defect. Published design values include strength ratios that are specific to species, stress type, nature of use, and lumber grade.

- Safety Factor: Wood's variability is accounted for in the 5% exclusion limit. The "safety factor" is intended to adjust properties for potential overload. The size of this safety factor depends on the predictability of behavior and the consequences of failure. There is no safety factor applied to compression across the grain or to modulus of elasticity. A timber "failure" due to either of these properties would only be a matter of unacceptable deflection -- not generally a collapse. Other safety factors range from about 0.8 for bending to 0.4 for shear, with shear behavior being more unpredictable than bending.

- Special Grading: Certain species have significantly variable properties, due to widely varying density within the species. The denser examples are stronger, and they can be graded accordingly. Since density variation within a species is largely a function of growth rate, the distinction is often done on the basis of growth ring, or "grain," spacing. The closer the ring spacing, the more dense will be the wood.

- Special Conditions: This catch-all factor covers a variety of issues. The depth of a bending member has an influence on its behavior, and is accounted for by normalizing to a twelve inch deep beam and readjusting for the actual depth. The modulus of elasticity is decreased to account for shear deflections in beams, which then can be conveniently neglected.

- Moisture Content: The test specifications call for specimens to be at equilibrium moisture content when tested. Lumber in service may have varying moisture content, which the engineer must take into account. As wood dries below the fiber saturation point, it becomes stronger, but the larger size members also split and are possibly weakened. Lumber is therefore divided into sizes larger than four inches wide and those which are narrower. The smaller lumber may be distinguished by as many as three moisture content ranges: above 19% (green), between 19% and 15%, and below 15%.

There are other factors, called "condition of use factors (CUF)," which are used to account for the special problems of designing connections in members which are expected to undergo changing moisture content and dimensions during their service life.

- Load Duration: A further manifestation of wood's complexity is its rheological or time-dependent stress-strain behavior. In general, wood's strength decreases as the duration of an applied load increases. Load durations range from "impact" loading to the permanent dead load.

The design life of timber structures is clearly longer than the duration of typical tests. Therefore, a factor is used to "normalize" the properties to a ten year duration. The engineer then corrects the allowable stresses to the actual expected load duration. Figure 5 illustrates the complete range of factors to be applied for load duration. The durations are cumulative over the life of the structure. The most commonly used load duration factors (LDF), used in all the codes, are given in table 2. These load duration factors apply to all connection design loads and allowable stress levels except modulus of elasticity and compression across the grain.

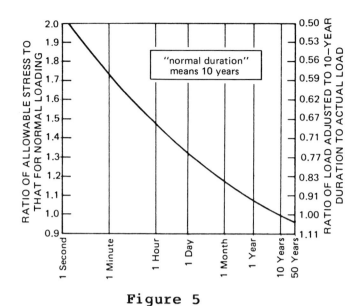

**Figure 5**

Relation of Strength to Duration of Load

**Table 2**

**Common Load Duration Factors**

Dead Load (50 years) ................. 0.90
Snow Load* (2 months) ................ 1.15
Roof Live Load* (7 days) ............. 1.25
Wind or Seismic Load* (1 day) ........ 1.33
Impact Load .......................... 2.00

  * inclusive of Dead Load

- <u>Load Combinations</u>: The duration of any given combination of load types is only that of the shortest duration load in the combination. That least value, therefore, establishes the load duration factor applied to the allowable stresses caused by the combination. Some authors make an issue of only applying the load duration factor to the allowable stresses and not to the load. But, the correct load duration factor must be applied to each of the possible design load combinations to determine which one is critical. The designer can apply the load duration factor to either load or allowable stress, as long as it is not applied to both load and allowable stress.

# Example 1

Determine the critical load combination on a roof rafter with the following loads: 20 psf dead load, 35 psf snow load, and 25 psf wind load.

To determine which combination of loads will control after modifying the allowable stresses for load duration, divide each combination by the load duration factor of the shortest duration load included in the combination. The maximum of these effective design loads establishes the controlling load combination.

| Load Combination | Total Load (psf) | Adjusted Load (psf) | Design Load (psf) |
|---|---|---|---|
| Dead Alone | 20 | 20/0.9 | 22.2 |
| Dead + Snow | 55 | 55/1.15 | 47.8 |
| Dead + Wind | 45 | 45/1.33 | 33.8 |
| Dead + Wind + Snow/2 | 62.5 | 62.5/1.33 | 47.0 |

The dead and snow load combination controls. The design could proceed with 47.8 psf and the table values for allowable stresses, or with 55 psf and allowable stresses increased by 15%. Dividing the snow load in half when combined with wind loads is a common building code recommendation. This may be nonconservative, particularly where unusual roof shapes could concentrate drifting snow, and should only be done with care and consideration.

## Full-Size Testing

Another method of establishing allowable stresses involves testing full-size members and modifying the determined strengths for design service conditions. The most common means of testing are measuring bending deflection or vibrating the piece to establish its frequency. These nondestructive tests provide data which can be correlated with the strength of the member.

Lumber that has been "stress-graded" by one of these methods will include an allowable bending stress in the grade stamp. For example, a 1600f-1.4E machine-stress-rated (MSR) board would have an allowable bending stress of 1600 psi and a modulus of elasticity (E) of $1.4 \times 10^6$ psi. The principal uses for stress-graded lumber in the United States are glue-laminated timbers and prefabricated trusses.

## 5. COMMERCIALLY AVAILABLE LUMBER AND ALLOWABLE DESIGN STRESSES

The vast majority of structural timber used in the United States is softwood. The major species are pines, firs, hemlocks, spruces, cedars, and redwood. Most of these species have several varieties, such as red pine and Ponderosa pine. When two or more similar species grow in the same area, they are often grouped together for marketing convenience. For example, "Hem-Fir" is a common pairing of hemlock and fir.

Trade associations developed around many of the species, and these associations usually wrote their own grading standards. The proliferation of different standards left a legacy of confusion that is still being settled. The various lumber categories still occasionally overlap, for example. Public Standard 20-70, administered by the American Lumber Standards Committee, has done much to standardize lumber grading across the nation. All the grading associations now use some form of a "select structural, #1, #2, #3, and utility" rankings to grade their lumber. A #2 Douglas fir 2x4 meets the same grading criteria as a #2 Southern pine 2x4. The allowable stresses and stiffnesses may not be equal, but their defects will be equivalent.

There are many grading standards that should not concern the engineer, such as ladder stock and pencil stock rules. This book uses the allowable stresses found in the Supplement to the National Design Specifications. A few of the species found in that publication are reproduced in appendix 2 of this book.

# Timber Design

## A. The Allowable Stress Tables

The tables of allowable stresses are reasonably easy to use once some of the terms are understood. The footnotes at the end of the tables can be as important as the tables themselves, but they are often ignored by the uninformed.

### Species/Variety

Determining the species of an old timber can be extremely difficult, even for experts in the field. Specifying a precise species for new construction can cause unnecessary expense and delays. Unless there is a real need for a specific species, it is best to determine what is commonly available, or simply to be conservative in species assumption.

### Moisture Content

Southern pine is one species further divided into categories depending on its moisture content when in use and while being manufactured. It is safe to assume that a timber protected from the elements will remain below a 15% moisture content, unless the timber is used in a cooling tower or other wet application.

### Grade and Size

Lumber is graded at the mill and grouped into lots. Grades better than #2 (e.g., #1, select structural, etc.) are generally in the minority. A common practice is to put the higher grades together as "#1 and better." Unless available space in the design is a serious problem, the engineer should check prices and availability before specifying lumber in the higher grades. It will often be cheaper to specify a larger size member of a lower grade.

### Sizes

Everyone seems to lament that a "two by four isn't one anymore." In other words, there is a difference between nominal and actual sizes. Lumber shrinks during manufacture by drying and being surfaced. A nominal 2x4 may start as nearly two inches by four inches, but it only has to be an inch and a half by three and a half to be called a "two by four."

Lumber's larger faces are known as sides, and the narrower faces are called edges. Lumber surfaced on one side and one edge is abbreviated as "1S1E," while the common lumber surfaced on all four sides is called "S4S."

The section properties of appendix 5 use the actual size of the member and neglect the standard corner rounding which is done to reduce splinters. A 4x2 is listed as well as a 2x4 to aid the engineer using the member in the flat orientation.

Older structures and some newer ones with heavy solid-sawn timbers may have full-size members in them. These members are generally rough sawn and unfinished. It is, of course, safe and conservative to assume that a given member size is a nominal one, and that the timber is not full-size unless specifically stated.

There are three basic size categories in the allowable stress tables: dimension lumber, posts and timbers, and beams and stringers. Dimension lumber is further broken down into size and use. Two of the dimension lumber subgroups deal with the same sizes -- maximum dimension four inches -- but are designated by either numbered grades or use grades. Which grade names are used depends on local supply and practice. The last dimension lumber subgroup contains those sizes with a dimension larger than four inches.

Posts and timbers are graded with the expectation, but not the requirement, that the members will be used as compression members. The minimum dimension of the post and timber size category is five inches, with no more than two inches difference in the two dimensions. Beams and stringers, on the other hand, also have a minimum dimension of five inches, but there must be more than a two inch difference in the two dimensions.

Glue-laminated members come in 5 1/8", 6 3/4", and 8 3/4" widths, laminated from 2x6's, 2x8's, and 2x10's respectively.  After gluing, the members are planed to the finish width.  The depth is a multiple of 1 1/2" or 3/4", depending on the number and thickness of the laminae.  Since the nominal size is equal to the actual size, section properties for glue-laminated members are readily calculated.

### B. Listed Properties

Once the correct row in the table has been found, it is a simple matter to read across to the required value in the appropriate column. The design properties contained in the National Design Specification tables duplicated in appendix 2 are:

<u>Extreme Fiber Stress in Bending</u> ($F_b$): This value is the allowable tension in a bending member cross-section. $F_b$ will be significantly higher than the allowable tension parallel to the grain ($F_t$) found in the adjacent column. The increase is due to the migrating neutral axis of timber in bending.

There is also a subcategory under $F_b$ called "repetitive use members", which has a higher $F_b$. The likelihood and consequences of one of a large number of interconnected members failing are so reduced that a lower safety factor has been used to derive the allowable bending stress. There are three criteria repetitive members must fulfill. There must be at least three members, they must not be more than 24 inches on center, and they must all be joined by a load-sharing member such as plywood or decking.

<u>Tension Parallel to the Grain</u> ($F_t$): This column contains the allowable tension in the longitudinal direction. Tension across the grain, either radial or tangential, is limited to one third the allowable horizontal shear ($F_v$). For some species, the allowable tension across the grain is further reduced to as low as fifteen pounds per square inch.

<u>Horizontal Shear</u> ($F_v$): Shear across the grain is seldom a concern. The allowable stresses listed in the column are shears in the

plane of the fibers, which tend to slide the wood cells by each other. Provision for increased allowable shear stresses can be found in the table footnotes. Allowable shear stresses are fairly low because of the susceptibility of an individual member to end splits that may open after grading.

<u>Compression Perpendicular to the Grain</u> ($F_{c\perp}$): In order to simplify design, there is no distinction made between compression in the radial and tangential directions. No safety factor is applied to the average of the test values for this property since overstressing will generally cause only minor localized crushing.

<u>Compression Parallel to the Grain</u> ($F_c$): This value is the one used to design columns and other compression members. The allowable stress is in the longitudinal direction. Long columns are designed for buckling, a phenomenon which is governed by stiffness (modulus of elasticity), not strength.

When the compression parallel to the grain is applied to the end of a member, the design values of appendix 1, End Grain in Bearing, apply. If the end grain bearing stress is larger than 75% of the allowable stress of appendix 1, a metal bearing plate is required to distribute the load.

<u>Modulus of Elasticity</u> (E): The number in this column is the longitudinal modulus of elasticity. The value is based on an average of test results, not a 5% exclusion limit, so a calculated deflection will also be an average expected value. The table modulus has been decreased by 10% from actual averages in order to allow for shear deflections in beams without actually having to calculate the shear component.

### C. Glue-Laminated Allowable Stress Tables

The allowable stress tables for glue-laminated members, in Appendix 2, read very much as do the lumber tables. The meaning of the various combination symbols is explained in the footnotes. The allowable stresses vary, depending on whether the applied load is mainly a transverse bending load or an axial one, and the bending axis orientation.

# 6. CONNECTIONS

## A. Introduction

In many ways, the specification of connections is the most demanding aspect of timber structural design. Timber members can be very strong, and yet they can be difficult to hold onto well enough to exploit that strength. The specification of a connection scheme requires attention to connection capacity, net sections, detailing, and maintaining certain minimum clearances. Maintaining edge and end distances in all the connected members may even force the engineer to use a larger member than is required by allowable stress levels within the member itself.

In roughly ascending order of holding capacity, the common mechanical fasteners used in the United States are nails and spikes, lag screws, bolts, shear plates, and split rings. Shear plates and split rings are manufactured by the Timber Engineering Company (TECO) and are commonly referred to as "connectors."[2] Adhesives are widely used to manufacture many timber structural elements, but adhesives have not found many applications in field-installed connections. With advances in technology and quality control, adhesives may be used more frequently.

## B. Fastener/Grain Relationship and Load Application Direction

The fastener's alignment with the wood fibers is of major importance. Fasteners which terminate within the member, such as nails and lag screws, can be installed either along the wood fibers (into the "end grain"), or across the wood fibers (into the "side grain"). The member which contains the point of these fasteners is the "holding member." The member through which the fastener passes is the "side member." Figure 6 illustrates these definitions.

---

[2] Timber Engineering Company, 5530 Wisconsin Avenue, Washington, D.C. 20015

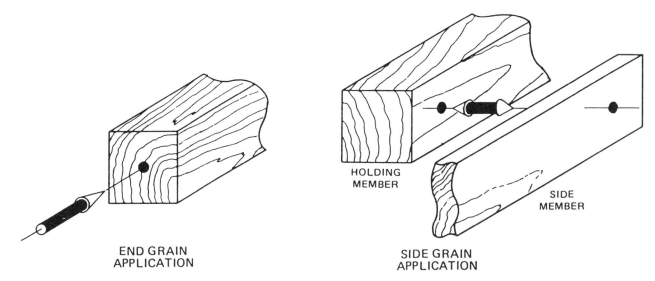

Figure 6
Direction of Fastener Installation

As shown in figure 7, the two basic load application directions are along the fastener axis (in tension or "withdrawal"), and normal to the fastener axis (in shear or "lateral load").

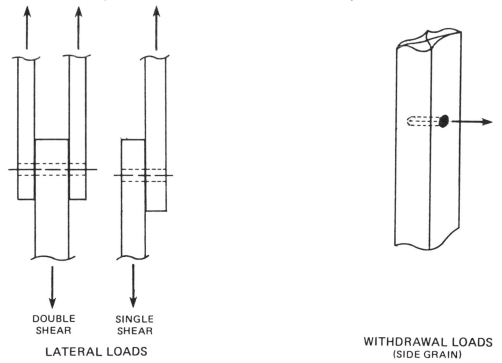

Figure 7
Types of Load Application

Lateral loads must be further distinguished by their orientation with the wood fibers or longitudinal axis of the timber member. Figure 8 shows a load applied parallel to the longitudinal axis which bears on the ends of the cells exposed in the hole through which the connector passes. The load capacity of such "parallel to grain" connections is significantly larger than those loaded "perpendicular to grain". Tension perpendicular to the grain, expected or unexpected, is the chief cause of connection failures.

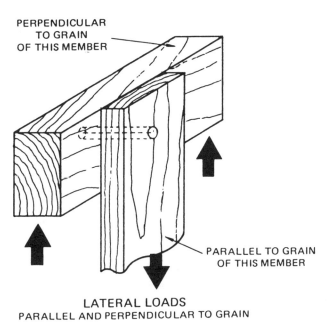

**Figure 8**

**Direction of Load Application Relative to Grain Direction**

The difference in capacity as related to orientation with the wood fibers is so profound that the engineer must investigate all the members in a multi-member bolted connection. When the load is applied at an angle which is neither parallel nor perpendicular to the grain, the Hankinson Formula (equation 2) is used to interpolate between the two limiting capacities.

$$F_n = \frac{F_c F_{c\perp}}{F_c \sin^2\theta + F_{c\perp} \cos^2\theta} \tag{2}$$

## Combinations of Load Types

A timber connection's response to lateral and withdrawal loadings is different enough that there is little interaction to consider. In a situation that combines both load types, the two applied loads can be compared independently with their respective capacities.

### Load Duration

The established practice is to assume that the capacity of all fasteners increases as loads are applied for shorter time periods. The load duration factors (LDF) of table 2 and figure 5 apply to all connections.

### Species Grouping

The capacity of any timber connection depends on the wood species involved. Through-bolts capacities are given in table 13 for a wide variety of species. For other fastener types, species are grouped together in capacity ranges. Withdrawal capacity is also a function of the specific gravity of the holding member. Table 3 is a list of species, their specific gravities, and the four common species groupings.

Glue-laminated members are classified, for connection design purposes, in footnote 14 of the glue-laminated allowable stress tables of appendix 2.

### Conditions of Use

Changing moisture conditions can have a tremendous influence on a connection's strength and endurance. The "condition of use" (CUF) factors of table 4 include capacity decreases as large as 75%.

### C. Nails and Spikes

Nails and spikes are the simplest and most common of the mechanical fasteners. They are often power-driven in both shop and field installations. The nail or spike must penetrate the holding member by a minimum amount in order to be fully effective. Table 5 lists the lengths and diameters of the most common sizes.

# Timber Design

## Table 3
## Species Groupings for Connection Design

| Group | Species of wood | Specific gravity** (G) |
|---|---|---|
| Group I | Ash, Commercial White | 0.62 |
| | Beech | 0.68 |
| | Birch, Sweet & Yellow | 0.66 |
| | Hickory & Pecan | 0.75 |
| | Maple, Black & Sugar | 0.66 |
| | Oak, Red & White | 0.67 |
| Group II | Douglas Fir - Larch*** | 0.51 |
| | Southern Pine | 0.55 |
| | Sweetgum & Tupelo | 0.54 |
| | Virginia Pine - Pond Pine | 0.54 |
| Group III | California Redwood | 0.42 |
| | Douglas Fir, South | 0.48 |
| | Eastern Hemlock | 0.43 |
| | Eastern Hemlock - Tamarack*** | 0.45 |
| | Eastern Softwoods | 0.42 |
| | Eastern Spruce | 0.43 |
| | Hem - Fir*** | 0.42 |
| | Lodgepole Pine | 0.44 |
| | Mountain Hemlock | 0.47 |
| | Mountain Hemlock - Hem Fir | 0.44 |
| | Northern Aspen | 0.42 |
| | Northern Pine | 0.46 |
| | Ponderosa Pine*** | 0.49 |
| | Ponderosa Pine-Sugar Pine | 0.42 |
| | Red Pine**** | 0.42 |
| | Sitka Spruce | 0.43 |
| | Southern Cypress | 0.48 |
| | Spruce-Pine-Fir | 0.42 |
| | Western Hemlock | 0.48 |
| | Yellow Poplar | 0.46 |
| Group IV | Aspen | 0.40 |
| | Balsam Fir | 0.38 |
| | Black Cottonwood | 0.33 |
| | California Redwood, Open grain | 0.37 |
| | Coast Sitka Spruce | 0.39 |
| | Coast Species | 0.39 |
| | Cottonwood, Eastern | 0.41 |
| | Eastern White Pine*** | 0.38 |
| | Eastern Woods | 0.38 |
| | Engelmann Spruce - Alpine Fir | 0.36 |
| | Idaho White Pine | 0.40 |
| | Northern Species | 0.35 |
| | Northern White Cedar | 0.31 |
| | West Coast Woods (Mixed Species) | 0.35 |
| | Western Cedars*** | 0.35 |
| | Western White Pine | 0.40 |
| | White Woods (Western Woods) | 0.35 |

**Based on weight and volume when oven-dry.
***Also applies when species name includes the designation "North".
****Applies when graded to NLGA rules.

By permission of NFPA

## Table 4
### Condition of Use Factors (CUF) for Connections

| Type of fastener | Condition of wood[1] | | Factor |
| --- | --- | --- | --- |
| | At time of fabrication | In service | |
| Timber connectors[2] | Dry | Dry | 1.0 |
| | Partially seasoned[3] | Dry | See Note 3 |
| | Wet | Dry | 0.8 |
| | Dry or wet | Partially seasoned or wet | 0.67 |
| Bolts or lag screws | Dry | Dry | 1.0 |
| | Partially seasoned[3] or wet | Dry | See Table 8.1C |
| | Dry or wet | Exposed to weather | 0.75 |
| | Dry or wet | Wet | 0.67 |
| Drift bolts or pins - Laterally loaded | Dry or wet | Dry | 1.0 |
| | Dry or wet | Partially seasoned or wet, or subject to wetting and drying | 0.70 |
| Wire nails and spikes | | | |
| —Withdrawal loads | Dry | Dry | 1.0 |
| | Partially seasoned or wet | Will remain wet | 1.0 |
| | Partially seasoned or wet | Dry | 0.25 |
| | Dry | Subject to wetting and drying | 0.25 |
| —Lateral loads | Dry | Dry | 1.0 |
| | Partially seasoned or wet | Dry or wet | 0.75 |
| | Dry | Partially seasoned or wet | 0.75 |
| Threaded, hardened steel nails | Dry or wet | Dry or wet | 1.0 |
| Wood screws | Dry or wet | Dry | 1.0 |
| | Dry or wet | Exposed to weather | 0.75 |
| | Dry or wet | Wet | 0.67 |
| Metal plate connectors | Dry | Dry | 1.0 |
| | Partially seasoned or wet | Dry or wet | 0.8 |

1. Condition of wood definitions applicable to fasteners are:

"Dry" wood has a moisture content of 19 percent or less.

"Wet" wood has a moisture content at or above the fiber saturation point (approximately 30 percent).

"Partially seasoned" wood, for the purposes of Table 8.1B, has a moisture content greater than 19 percent but less than the fiber saturation point (approximately 30 percent).

"Exposed to weather" implies that the wood may vary in moisture content from dry to partially seasoned, but is not expected to reach the fiber saturation point at times when the joint is under full design load.

"Subject to wetting and drying" implies that the wood may vary in moisture content from dry to partially seasoned or wet, or vice versa, with consequent effects on the tightness of the joint.

2. For timber connectors, moisture content limitations apply to a depth of 3/4 inch from the surface of the wood.

3. When timber connectors, bolts or laterally loaded lag screws are installed in wood that is partially seasoned at the time of fabrication but that will be dry before full design load is applied, proportional intermediate values may be used.

## Table 5
### Common Nail and Spike Dimensions

| Pennyweight | Length Inches | Wire diameter, inches | | | |
| --- | --- | --- | --- | --- | --- |
| | | Box nails | Common wire nails | Threaded hardened-steel nails | Common wire spikes |
| 6d | 2 | 0.099 | 0.113 | 0.120 | — |
| 8d | 2½ | 0.113 | 0.131 | 0.120 | — |
| 10d | 3 | 0.128 | 0.148 | 0.135 | 0.192 |
| 12d | 3¼ | 0.128 | 0.148 | 0.135 | 0.192 |
| 16d | 3½ | 0.135 | 0.162 | 0.148 | 0.207 |
| 20d | 4 | 0.148 | 0.192 | 0.177 | 0.225 |
| 30d | 4½ | 0.148 | 0.207 | 0.177 | 0.244 |
| 40d | 5 | 0.162 | 0.225 | 0.177 | 0.263 |
| 50d | 5½ | — | 0.244 | 0.177 | 0.283 |
| 60d | 6 | — | 0.263 | 0.177 | 0.283 |
| 70d | 7 | — | — | 0.207 | — |
| 80d | 8 | — | — | 0.207 | — |
| 90d | 9 | — | — | 0.207 | — |
| 5/16 | 7 | — | — | — | 0.312 |
| 3/8 | 8½ | — | — | — | 0.375 |

By permission of NFPA

## Connection Capacity -- Multiple Nails and Spikes

Whether the nailed connection is loaded laterally or in withdrawal, the connection capacity is a simple multiple of the number of nails and their individual capacities. The only limit on this linear increase in strength is reached when another nail would split the member.

## Lateral Loads

Figure 9 shows a typical load-deformation curve for a laterally loaded, nailed connection.

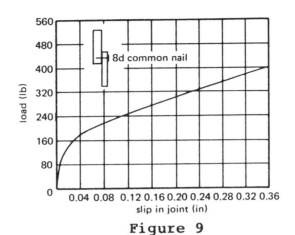

**Figure 9**

**Load-Deformation Curve for Typical Laterally Loaded Nail**

Clearly, the deformations become unacceptably high long before actual failure occurs. A 0.015" deflection is an accepted proportional limit for the connections, and is used as the failure criterion. The loads corresponding to this deformation are so low that the load direction makes no difference in the nail capacity. Therefore, there is no distinction drawn between loads parallel and perpendicular to the grain with nails, nor any application for the Hankinson formula.

- End Grain: Nails in the end grain of the holding member have only 2/3 the lateral load capacity of the same nail penetrating the same distance into the sidegrain of the same species.

- Penetration: The penetration of the nail into the holding member must be greater than a specific value for the nail to develop its full lateral load capacity. The minimum required full-strength penetration is a function of the nail diameter and the density of the holding member species. For the species of Group 1 in table 3, the full-strength penetration in the holding member is 10 nail diameters. Similarly, the minimum penetration is 11 diameters for Group 2 species, 13 diameters for Group 3 species, and 14 diameters for Group 4 species. These values are included in table 6, for convenience.

The minimum allowable penetration is 1/3 the full-strength penetration. For any nail penetration less than full-strength, the capacity can be determined by a simple interpolation between zero capacity at no penetration and full strength at the specified penetration.

- **Metal Side Plates:** If the nails attach an adequately stiff and strong metal sideplate to the timber holding member, the allowable loads can be increased 25%.

- **Different Species:** If two different species are nailed together, the lesser design value controls regardless of whether it is the holding or side member.

- **Toe-nailed Connections:** Some nailed connections must have the nails installed as "toe-nails" because of clearance restrictions or construction sequence. The lateral capacity of these connections is 5/6 that of the corresponding lateral load nail-species combination. The withdrawal capacity of toe-nailed connections is 2/3 of the corresponding straight nailed connection.

### Allowable Lateral Loads

The design values for laterally loaded nails and spikes are found in table 6. These are normal load duration values, specific for species group and nail size, and based on full-strength penetration.

### Withdrawal Loads

The design withdrawal capacity of nails and spikes is so low and so susceptible to weakening with changed moisture content, that consideration should be given to either other fastener types or revising the connection layout to one with lateral loading. There is no dependable capacity if the nail is in the end grain of the holding member.

If a nailed connection must be loaded in withdrawal, the capacity is a multiple of the total number of inches of nail penetration into the holding member and the allowable withdrawal load per inch of that nail

# Table 6
## Design Values -- Nails and Spikes in Lateral Loads

Normal load duration

Design values for lateral loads (single shear) for nails and spikes penetrating not less than 10 diameters in Group I species, 11 diameters in Group II species, 13 in diameters in Group III species, and 14 diameters in Group IV species, into the member holding the point. Nail size in pennyweight. Diameters and lengths in inches. Loads in pounds.

### BOX NAILS

| | | | | | | | | |
|---|---|---|---|---|---|---|---|---|
| Penny weight | 6d | 8d | 10d | 12d | 16d | 20d | 30d | 40d |
| Length | 2 | 2½ | 3 | 3¼ | 3½ | 4 | 4½ | 5 |
| Diameter | 0.099 | 0.113 | 0.128 | 0.128 | 0.135 | 0.148 | 0.148 | 0.162 |
| 10 Diameters | 0.99 | 1.13 | 1.28 | 1.28 | 1.35 | 1.48 | 1.48 | 1.62 |
| 11 Diameters | 1.09 | 1.24 | 1.41 | 1.41 | 1.49 | 1.63 | 1.63 | 1.78 |
| 13 Diameters | 1.29 | 1.47 | 1.66 | 1.66 | 1.76 | 1.92 | 1.92 | 2.11 |
| 14 Diameters | 1.39 | 1.58 | 1.79 | 1.79 | 1.89 | 2.07 | 2.07 | 2.27 |
| Species group I | 64 | 77 | 93 | 93 | 101 | 116 | 116 | 133 |
| Species group II | 51 | 63 | 76 | 76 | 82 | 94 | 94 | 108 |
| Species group III | 42 | 51 | 62 | 62 | 67 | 77 | 77 | 88 |
| Species group IV | 34 | 41 | 49 | 49 | 54 | 61 | 61 | 70 |

### COMMON WIRE NAILS

| | | | | | | | | | | |
|---|---|---|---|---|---|---|---|---|---|---|
| Penny weight | 6d | 8d | 10d | 12d | 16d | 20d | 30d | 40d | 50d | 60d |
| Length | 2 | 2½ | 3 | 3¼ | 3½ | 4 | 4½ | 5 | 5½ | 6 |
| Diameter | 0.113 | 0.131 | 0.148 | 0.148 | 0.162 | 0.192 | 0.207 | 0.225 | 0.244 | 0.263 |
| 10 Diameters | 1.13 | 1.31 | 1.48 | 1.48 | 1.62 | 1.92 | 2.07 | 2.25 | 2.44 | 2.63 |
| 11 Diameters | 1.24 | 1.44 | 1.63 | 1.63 | 1.78 | 2.11 | 2.28 | 2.48 | 2.68 | 2.89 |
| 13 Diameters | 1.47 | 1.70 | 1.92 | 1.92 | 2.11 | 2.50 | 2.69 | 2.93 | 3.17 | 3.42 |
| 14 Diameters | 1.58 | 1.83 | 2.07 | 2.07 | 2.27 | 2.69 | 2.90 | 3.15 | 3.42 | 3.68 |
| Species group I | 77 | 97 | 116 | 116 | 133 | 172 | 192 | 218 | 246 | 275 |
| Species group II | 63 | 78 | 94 | 94 | 108 | 139 | 155 | 176 | 199 | 223 |
| Species group III | 51 | 64 | 77 | 77 | 88 | 114 | 127 | 144 | 163 | 182 |
| Species group IV | 41 | 51 | 61 | 61 | 70 | 91 | 102 | 115 | 130 | 146 |

### THREADED HARDENED STEEL NAILS AND SPIKES

| | | | | | | | | | | | | | |
|---|---|---|---|---|---|---|---|---|---|---|---|---|---|
| Penny weight | 6d | 8d | 10d | 12d | 16d | 20d | 30d | 40d | 50d | 60d | 70d | 80d | 90d |
| Length | 2 | 2½ | 3 | 3¼ | 3½ | 4 | 4½ | 5 | 5½ | 6 | 7 | 8 | 9 |
| Diameter | 0.120 | 0.120 | 0.135 | 0.135 | 0.148 | 0.177 | 0.177 | 0.177 | 0.177 | 0.177 | 0.207 | 0.207 | 0.207 |
| 10 Diameters | 1.20 | 1.20 | 1.35 | 1.35 | 1.48 | 1.77 | 1.77 | 1.77 | 1.77 | 1.77 | 2.07 | 2.07 | 2.07 |
| 11 Diameters | 1.32 | 1.32 | 1.49 | 1.49 | 1.63 | 1.95 | 1.95 | 1.95 | 1.95 | 1.95 | 2.28 | 2.28 | 2.28 |
| 13 Diameters | 1.56 | 1.56 | 1.76 | 1.76 | 1.92 | 2.30 | 2.30 | 2.30 | 2.30 | 2.30 | 2.69 | 2.69 | 2.69 |
| 14 Diameters | 1.68 | 1.68 | 1.89 | 1.89 | 2.07 | 2.48 | 2.48 | 2.48 | 2.48 | 2.48 | 2.90 | 2.90 | 2.90 |
| Species group I | 77 | 97 | 116 | 116 | 133 | 172 | 172 | 172 | 172 | 172 | 218 | 218 | 218 |
| Species group II | 63 | 78 | 94 | 94 | 108 | 139 | 139 | 139 | 139 | 139 | 176 | 176 | 176 |
| Species group III | 51 | 64 | 77 | 77 | 88 | 114 | 114 | 114 | 114 | 114 | 144 | 144 | 144 |
| Species group IV | 41 | 51 | 61 | 61 | 70 | 91 | 91 | 91 | 91 | 91 | 115 | 115 | 115 |

### COMMON WIRE SPIKES

| | 10d | 12d | 16d | 20d | 30d | 40d | 50d | 60d | 5/16" | 3/8" |
|---|---|---|---|---|---|---|---|---|---|---|
| Penny weight | 10d | 12d | 16d | 20d | 30d | 40d | 50d | 60d | 5/16" | 3/8" |
| Length | 3 | 3¼ | 3½ | 4 | 4½ | 5 | 5½ | 6 | 7 | 8½ |
| Diameter | 0.192 | 0.192 | 0.207 | 0.225 | 0.244 | 0.263 | 0.283 | 0.283 | 0.312 | 0.375 |
| 10 Diameters | 1.92 | 1.92 | 2.07 | 2.25 | 2.44 | 2.63 | 2.83 | 2.83 | 3.12 | 3.75 |
| 11 Diameters | 2.11 | 2.11 | 2.28 | 2.48 | 2.68 | 2.89 | 3.11 | 3.11 | 3.43 | 4.13 |
| 13 Diameters | 2.50 | 2.50 | 2.69 | 2.93 | 3.17 | 3.42 | 3.68 | 3.68 | 4.06 | 4.88 |
| 14 Diameters | 2.69 | 2.69 | 2.90 | 3.15 | 3.42 | 3.68 | 3.96 | 3.96 | 4.37 | 5.25 |
| Species group I | 172 | 172 | 192 | 218 | 246 | 275 | 307 | 307 | 356 | 468 |
| Species group II | 139 | 139 | 155 | 176 | 199 | 223 | 248 | 248 | 288 | 379 |
| Species group III | 114 | 114 | 127 | 144 | 163 | 182 | 203 | 203 | 235 | 310 |
| Species group IV | 91 | 91 | 102 | 115 | 130 | 146 | 163 | 163 | 188 | 248 |

By permission of NFPA

## Table 7

### Allowable Withdrawal Loads -- Nails and Spikes
Normal load duration

Design values in withdrawal in pounds per inch of penetration into side grain of member holding point.

$d$ = pennyweight of nail or spike. $G$ = specific gravity of the wood, based on weight and volume when oven-dry.

| Specific gravity $G$ | | Size of common nail | | | | | | | | | | Size of threaded nail* | | | | | | |
|---|---|---|---|---|---|---|---|---|---|---|---|---|---|---|---|---|---|---|
| | Penny wt. Diam. | 6d 0.113 | 8d 0.131 | 10d 0.148 | 12d 0.148 | 16d 0.162 | 20d 0.192 | 30d 0.207 | 40d 0.225 | 50d 0.244 | 60d 0.263 | 30d 0.177 | 40d 0.177 | 50d 0.177 | 60d 0.177 | 70d 0.207 | 80d 0.207 | 90d 0.207 |
| 0.75 | | 76 | 88 | 99 | 99 | 109 | 129 | 139 | 151 | 164 | 177 | 129 | 129 | 129 | 129 | 151 | 151 | 151 |
| 0.68 | | 59 | 69 | 78 | 78 | 85 | 101 | 109 | 118 | 128 | 138 | 101 | 101 | 101 | 101 | 118 | 118 | 118 |
| 0.67 | | 57 | 66 | 75 | 75 | 82 | 97 | 105 | 114 | 124 | 133 | 97 | 97 | 97 | 97 | 114 | 114 | 114 |
| 0.66 | | 55 | 64 | 72 | 72 | 79 | 94 | 101 | 110 | 119 | 128 | 94 | 94 | 94 | 94 | 110 | 110 | 110 |
| 0.62 | | 47 | 55 | 62 | 62 | 68 | 80 | 86 | 94 | 102 | 110 | 80 | 80 | 80 | 80 | 94 | 94 | 94 |
| 0.55 | | 35 | 41 | 46 | 46 | 50 | 59 | 64 | 70 | 76 | 81 | 59 | 59 | 59 | 59 | 70 | 70 | 70 |
| 0.54 | | 33 | 39 | 44 | 44 | 48 | 57 | 61 | 67 | 72 | 78 | 57 | 57 | 57 | 57 | 67 | 67 | 67 |
| 0.51 | | 29 | 34 | 38 | 38 | 42 | 49 | 53 | 58 | 63 | 67 | 49 | 49 | 49 | 49 | 58 | 58 | 58 |
| 0.49 | | 26 | 30 | 34 | 34 | 38 | 45 | 48 | 52 | 57 | 61 | 45 | 45 | 45 | 45 | 52 | 52 | 52 |
| 0.48 | | 25 | 29 | 33 | 33 | 36 | 42 | 46 | 50 | 54 | 58 | 42 | 42 | 42 | 42 | 50 | 50 | 50 |
| 0.47 | | 24 | 27 | 31 | 31 | 34 | 40 | 43 | 47 | 51 | 55 | 40 | 40 | 40 | 40 | 47 | 47 | 47 |
| 0.46 | | 22 | 26 | 29 | 29 | 32 | 38 | 41 | 45 | 48 | 52 | 38 | 38 | 38 | 38 | 45 | 45 | 45 |
| 0.45 | | 21 | 25 | 28 | 28 | 30 | 36 | 39 | 42 | 46 | 49 | 36 | 36 | 36 | 36 | 42 | 42 | 42 |
| 0.44 | | 20 | 23 | 26 | 26 | 29 | 34 | 37 | 40 | 43 | 47 | 34 | 34 | 34 | 34 | 40 | 40 | 40 |
| 0.43 | | 19 | 22 | 25 | 25 | 27 | 32 | 35 | 38 | 41 | 44 | 32 | 32 | 32 | 32 | 38 | 38 | 38 |
| 0.42 | | 18 | 21 | 23 | 23 | 26 | 30 | 33 | 35 | 38 | 41 | 30 | 30 | 30 | 30 | 35 | 35 | 35 |
| 0.41 | | 17 | 19 | 22 | 22 | 24 | 29 | 31 | 33 | 36 | 39 | 29 | 29 | 29 | 29 | 33 | 33 | 33 |
| 0.40 | | 16 | 18 | 21 | 21 | 23 | 27 | 29 | 31 | 34 | 37 | 27 | 27 | 27 | 27 | 31 | 31 | 31 |
| 0.39 | | 15 | 17 | 19 | 19 | 21 | 25 | 27 | 29 | 32 | 34 | 25 | 25 | 25 | 25 | 29 | 29 | 29 |
| 0.38 | | 14 | 16 | 18 | 18 | 20 | 24 | 25 | 28 | 30 | 32 | 24 | 24 | 24 | 24 | 28 | 28 | 28 |
| 0.37 | | 13 | 15 | 17 | 17 | 19 | 22 | 24 | 26 | 28 | 30 | 22 | 22 | 22 | 22 | 26 | 26 | 26 |
| 0.36 | | 12 | 14 | 16 | 16 | 17 | 21 | 22 | 24 | 26 | 28 | 21 | 21 | 21 | 21 | 24 | 24 | 24 |
| 0.35 | | 11 | 13 | 15 | 15 | 16 | 19 | 21 | 23 | 24 | 26 | 19 | 19 | 19 | 19 | 23 | 23 | 23 |
| 0.33 | | 10 | 11 | 13 | 13 | 14 | 17 | 18 | 19 | 21 | 23 | 17 | 17 | 17 | 17 | 19 | 19 | 19 |
| 0.31 | | 8 | 10 | 11 | 11 | 12 | 14 | 15 | 17 | 18 | 19 | 14 | 14 | 14 | 14 | 17 | 17 | 17 |

*Loads for threaded, hardened steel nails, in 6d to 20d sizes, are the same as for common nails.

| Specific gravity $G$ | | Size of box nail | | | | | | | | Size of common spike | | | | | | | | |
|---|---|---|---|---|---|---|---|---|---|---|---|---|---|---|---|---|---|---|
| | Penny wt. Diam. | 6d 0.099 | 8d 0.113 | 10d 0.128 | 12d 0.128 | 16d 0.135 | 20d 0.148 | 30d 0.148 | 40d 0.162 | 10d 0.192 | 12d 0.192 | 16d 0.207 | 20d 0.225 | 30d 0.244 | 40d 0.263 | 50d 0.283 | 60d 0.283 | 5/16" 0.312 | 3/8" 0.375 |
| 0.75 | | 67 | 76 | 86 | 86 | 91 | 99 | 99 | 109 | 129 | 129 | 139 | 151 | 164 | 177 | 190 | 190 | 210 | 252 |
| 0.68 | | 52 | 59 | 67 | 67 | 71 | 78 | 78 | 85 | 101 | 101 | 109 | 118 | 128 | 138 | 149 | 149 | 164 | 197 |
| 0.67 | | 50 | 57 | 65 | 65 | 68 | 75 | 75 | 82 | 97 | 97 | 105 | 114 | 124 | 133 | 144 | 144 | 158 | 190 |
| 0.66 | | 48 | 55 | 63 | 63 | 66 | 72 | 72 | 79 | 94 | 94 | 101 | 110 | 119 | 128 | 138 | 138 | 152 | 183 |
| 0.62 | | 41 | 47 | 53 | 53 | 56 | 62 | 62 | 68 | 80 | 80 | 86 | 94 | 102 | 110 | 118 | 118 | 130 | 157 |
| 0.55 | | 31 | 35 | 40 | 40 | 42 | 46 | 46 | 50 | 59 | 59 | 64 | 70 | 76 | 81 | 88 | 88 | 97 | 116 |
| 0.54 | | 29 | 33 | 38 | 38 | 40 | 44 | 44 | 48 | 57 | 57 | 61 | 67 | 72 | 78 | 84 | 84 | 92 | 111 |
| 0.51 | | 25 | 29 | 33 | 33 | 35 | 38 | 38 | 42 | 49 | 49 | 53 | 58 | 63 | 67 | 73 | 73 | 80 | 96 |
| 0.49 | | 23 | 26 | 30 | 30 | 31 | 34 | 34 | 38 | 45 | 45 | 48 | 52 | 57 | 61 | 66 | 66 | 72 | 87 |
| 0.48 | | 22 | 25 | 28 | 28 | 30 | 33 | 33 | 36 | 42 | 42 | 46 | 50 | 54 | 58 | 62 | 62 | 69 | 83 |
| 0.47 | | 21 | 24 | 27 | 27 | 28 | 31 | 31 | 34 | 40 | 40 | 43 | 47 | 51 | 55 | 59 | 59 | 65 | 78 |
| 0.46 | | 20 | 22 | 25 | 25 | 27 | 29 | 29 | 32 | 38 | 38 | 41 | 45 | 48 | 52 | 56 | 56 | 62 | 74 |
| 0.45 | | 19 | 21 | 24 | 24 | 25 | 28 | 28 | 30 | 36 | 36 | 39 | 42 | 46 | 49 | 53 | 53 | 58 | 70 |
| 0.44 | | 18 | 20 | 23 | 23 | 24 | 26 | 26 | 29 | 34 | 34 | 37 | 40 | 43 | 47 | 50 | 50 | 55 | 66 |
| 0.43 | | 17 | 19 | 21 | 21 | 23 | 25 | 25 | 27 | 32 | 32 | 35 | 38 | 41 | 44 | 47 | 47 | 52 | 63 |
| 0.42 | | 16 | 18 | 20 | 20 | 21 | 23 | 23 | 26 | 30 | 30 | 33 | 35 | 38 | 41 | 45 | 45 | 49 | 59 |
| 0.41 | | 15 | 17 | 19 | 19 | 20 | 22 | 22 | 24 | 29 | 29 | 31 | 33 | 36 | 39 | 42 | 42 | 46 | 56 |
| 0.40 | | 14 | 16 | 18 | 18 | 19 | 21 | 21 | 23 | 27 | 27 | 29 | 31 | 34 | 37 | 40 | 40 | 44 | 52 |
| 0.39 | | 13 | 15 | 17 | 17 | 18 | 19 | 19 | 21 | 25 | 25 | 27 | 29 | 32 | 34 | 37 | 37 | 41 | 49 |
| 0.38 | | 12 | 14 | 16 | 16 | 17 | 18 | 18 | 20 | 24 | 24 | 25 | 28 | 30 | 32 | 35 | 35 | 38 | 46 |
| 0.37 | | 11 | 13 | 15 | 15 | 16 | 17 | 17 | 19 | 22 | 22 | 24 | 26 | 28 | 30 | 33 | 33 | 36 | 43 |
| 0.36 | | 11 | 12 | 14 | 14 | 14 | 16 | 16 | 17 | 21 | 21 | 22 | 26 | 26 | 28 | 30 | 30 | 33 | 40 |
| 0.35 | | 10 | 11 | 13 | 13 | 14 | 15 | 15 | 16 | 19 | 19 | 21 | 23 | 24 | 26 | 28 | 28 | 31 | 38 |
| 0.33 | | 9 | 10 | 11 | 11 | 12 | 13 | 13 | 14 | 17 | 17 | 18 | 19 | 21 | 23 | 24 | 24 | 27 | 32 |
| 0.31 | | 7 | 8 | 9 | 9 | 10 | 11 | 11 | 12 | 14 | 14 | 15 | 17 | 18 | 19 | 21 | 21 | 23 | 28 |

By permission of NFPA

type in the holding member's species. The allowable withdrawal load per inch of penetration, found in table 7, is a function of the holding member's specific gravity (see table 3) and of the nail type and size.

- <u>Modifications to Allowable Withdrawal Loads</u>: The withdrawal capacity is subject to the standard load duration factors and the condition of use factors of table 4. There is no increase in withdrawal capacity with metal side plates.

## Example 2

What is the allowable combined snow and dead load on the exterior, overhead trellis slat shown? All timber is treated with preservative.

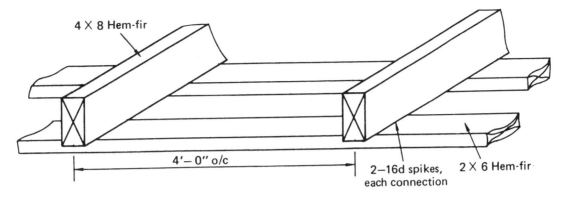

The solution procedure is to establish the total penetration of the nails and multiply by the allowable withdrawal load per inch of penetration to calculate the connection capacity. Then, convert this capacity to a snow load on the slats.

Subtracting the 2x6 hem-fir thickness, the actual nail penetration is 3.5" - 1.5" = 2". Table 5 yields the 3 1/2" nail length.

From table 3, hem-fir is in group III, and it has a specific gravity of G = .42. From table 7 for 16d spikes at G = .42, the withdrawal design value is 33 lb/in of penetration.

From table 4, the condition of use factor is CUF = .25. The load duration factor for snow loading is LDF = 1.15, from table 2.

The capacity per connection is

(2) nails (2) in/nail (33) lb/in (1.15)(.25) = 40.0 lb/connection

The allowable combined snow and dead load is

$$\frac{(40.0) \text{ lb/connection}}{(4) \text{ ft } (5.5"/12) \text{ ft}} = 22 \text{ psf}$$

Note: The pressure treatment has no effect on capacity.

## Example 3

Investigate the connection capacity at point A at the top of the backboard. Assume a 230 pound player occasionally and briefly hangs on the rim.

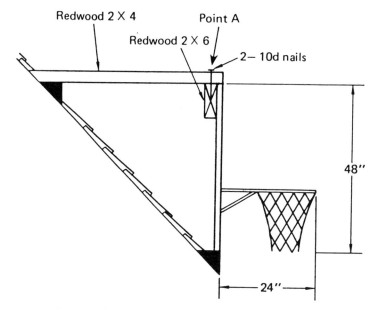

The solution procedure is to determine the load applied on the connection, check the actual penetration, and compare the capacity to the applied load.

The design load for connection at A is (230) lb (2 ft/4 ft) = 115 lb

From table 4, CUF = .75. From table 2, LDF = 2.0 (impact load). From table 3, redwood is in group III. Group III species have a full strength penetration of 13 diameters, or 1.92" with 10d nails (table 6).

A 10d nail is 3" long (table 5). The actual penetration of the nails into the 2x6 is 3" - 1.5" = 1.5". This actual penetration is greater than the absolute minimum required, one-third of the full strength, or 1.92"/3 = .64". A simple interpolation factor of 1.5"/1.92" will adjust the nail capacity for the actual penetration.

Table 6 gives the capacity of one 10d nail in a group III species as 77 lb/nail.

The connection capacity is

   (2) nails (77) lb/nail (2.0) (.75) (1.5"/1.92") = 180 lb

The 180 lb capacity exceeds the 115 lb design load, therefore the design is satisfactory.

# Example 4

What is the design capacity of the X-brace in the framing of a garage wall resisting wind load?

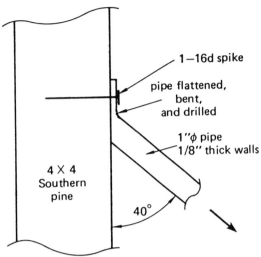

First, check the lateral capacity of the nail as it limits the lateral component of the load. Then, check the withdrawal capacity of the nail as it limits the withdrawal component of the load. The lower value will control the connection capacity.

Table 4 gives the CUF = 1.0. Wind load means a LDF = 1.33 (table 2). The flattened pipe acts as a metal side plate and increases only the lateral nail capacity by 25%. Southern pine is a group II species with a specific gravity, G = .55 (table 3).

A 16d spike is 3.5" long (table 5). After subtracting two pipe wall thicknesses, the penetration is 3.5" - (2)(.125") = 3.25". Table 7 gives a withdrawal capacity of 64 lb/in. The sine of 40° is used to determine the withdrawal component of the applied load.

The connection capacity, as limited by the withdrawal capacity, is

$$(64) \text{ lb/in } (3.25) \text{ in } (1.0)(1.33) / \sin 40° = 430 \text{ lb}$$

Group II species have a full lateral strength penetration of 11 diameters, or 2.28" with 16d spikes (table 6). The actual 3.25" penetration exceeds the required 2.28". There is no increase in the lateral capacity over the full strength value for the extra penetration. Table 6 gives a capacity of 155 lb for a 16d spike. The cosine of 40° is used to convert the lateral component to an applied load.

The connection capacity, as limited by the lateral capacity is

$$(155) \text{ lb } (1.0)(1.33)(1.25)/\cos 40° = 336 \text{ lb}$$

Therefore, lateral capacity limits design to 336 lb.

There is no interaction considered between lateral and withdrawal loads.

### D. Bolts, Lag Screws, Shear Plates, and Split Rings

When the loads and members are too large for nails, a heavier mechanical fastener must be used. A bolt in timber acts much as it does in metal, except that all of the load is transferred in bearing, with no clamping or friction effects. A lag screw is a large wood screw used to resist lateral and/or withdrawal loads when access to the far side of the member is restricted by its size or the erection sequence. Shear plates and split rings are specialized heavy-duty connectors which require special cutting heads for installation.

#### Capacity Reduction with Multiple Fasteners

Unlike nailed connections, the shear capacity of a connection using any of the other mechanical fasteners is not a simple multiple of the number of fasteners and their individual capacities. Because of unequal load sharing among these stiffer fasteners, capacities are reduced on the basis of connection layout and the side member material and relative size.

To determine the multiple fastener reduction factor, K, the number of fasteners in any row must first be determined. A row is defined as two or more fasteners lying in a line parallel to the direction of the applied load. Adjacent rows of staggered fasteners are considered to act as a single row if the distance between the rows is less than one-fourth the least distance, along the load axis, between fasteners in the adjacent rows.

# Timber Design

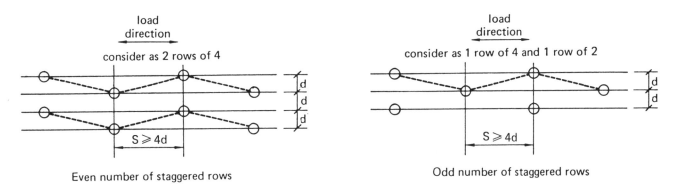

**Figure 10
Number of Fasteners in a Row**

Once the number of fasteners per row is known, the modification factor is found in table 8A or table 8B, depending on whether the side member is timber or metal. In either table, the relative and absolute sizes of the side and main members determines the modification factor applied to the sum of the individual capacities of the fasteners.

The multiple fastener factor, K, usually has little effect on connections with fewer than five fasteners in a row, but in the bigger connections the effect can be profound. The effect of this factor can be so large as to make an additional fastener in a row actually decrease the overall connection design capacity. Careful attention should be paid to connection layout in very heavy splices.

**Table 8A
Modification Factor for Number of Fasteners in Row, K
Wood Side Plates**

| $A_1/A_2$ | $A_1$ (in$^2$)† | Number of fasteners in a row | | | | | | | | | | |
|---|---|---|---|---|---|---|---|---|---|---|---|---|
| | | 2 | 3 | 4 | 5 | 6 | 7 | 8 | 9 | 10 | 11 | 12 |
| 0.5 • ‡ | <12 | 1.00 | 0.92 | 0.84 | 0.76 | 0.68 | 0.61 | 0.55 | 0.49 | 0.43 | 0.38 | 0.34 |
| | 12 – 19 | 1.00 | 0.95 | 0.88 | 0.82 | 0.75 | 0.68 | 0.62 | 0.57 | 0.52 | 0.48 | 0.43 |
| | >19 – 28 | 1.00 | 0.97 | 0.93 | 0.88 | 0.82 | 0.77 | 0.71 | 0.67 | 0.63 | 0.59 | 0.55 |
| | >28 – 40 | 1.00 | 0.98 | 0.96 | 0.92 | 0.87 | 0.83 | 0.79 | 0.75 | 0.71 | 0.69 | 0.66 |
| | >40 – 64 | 1.00 | 1.00 | 0.97 | 0.94 | 0.90 | 0.86 | 0.83 | 0.79 | 0.76 | 0.74 | 0.72 |
| | >64 | 1.00 | 1.00 | 0.98 | 0.95 | 0.91 | 0.88 | 0.85 | 0.82 | 0.80 | 0.78 | 0.76 |
| 1.0 • ‡ | <12 | 1.00 | 0.97 | 0.92 | 0.85 | 0.78 | 0.71 | 0.65 | 0.59 | 0.54 | 0.49 | 0.44 |
| | 12 – 19 | 1.00 | 0.98 | 0.94 | 0.89 | 0.84 | 0.78 | 0.72 | 0.66 | 0.61 | 0.56 | 0.51 |
| | >19 – 28 | 1.00 | 1.00 | 0.97 | 0.93 | 0.89 | 0.85 | 0.80 | 0.76 | 0.72 | 0.68 | 0.64 |
| | >28 – 40 | 1.00 | 1.00 | 0.99 | 0.96 | 0.92 | 0.89 | 0.86 | 0.83 | 0.80 | 0.78 | 0.75 |
| | >40 – 64 | 1.00 | 1.00 | 1.00 | 0.97 | 0.94 | 0.91 | 0.88 | 0.85 | 0.84 | 0.82 | 0.80 |
| | >64 | 1.00 | 1.00 | 1.00 | 0.99 | 0.96 | 0.93 | 0.91 | 0.88 | 0.87 | 0.86 | 0.85 |

Notes: 1. $A_1$ = cross-sectional area of main member(s) before boring or grooving.
2. $A_2$ = sum of the cross-sectional areas of side members before boring or grooving.
*When $A_1/A_2$ exceeds 1.0, use $A_2/A_1$.
†When $A_1/A_2$ exceeds 1.0, use $A_2$ instead of $A_1$.
‡For $A_1/A_2$ between 0 and 1.0, interpolate or extrapolate from the tabulated values

By permission of NFPA

## Table 8B
### Modification Factor for Number of Fasteners in Row, K
### Metal Side Plates

| $A_1/A_2$ | $A_1$ (in$^2$) | \multicolumn{11}{c|}{Number of fasteners in a row} |
|---|---|---|---|---|---|---|---|---|---|---|---|---|

| $A_1/A_2$ | $A_1$ (in$^2$) | 2 | 3 | 4 | 5 | 6 | 7 | 8 | 9 | 10 | 11 | 12 |
|---|---|---|---|---|---|---|---|---|---|---|---|---|
| 2–12 | 5 – 8 | 1.00 | 0.78 | 0.64 | 0.54 | 0.46 | 0.40 | 0.35 | 0.30 | 0.25 | 0.20 | 0.15 |
|  | 9 – 16 | 1.00 | 0.85 | 0.73 | 0.63 | 0.54 | 0.48 | 0.42 | 0.38 | 0.34 | 0.30 | 0.26 |
|  | 17 – 24 | 1.00 | 0.91 | 0.83 | 0.74 | 0.66 | 0.59 | 0.53 | 0.48 | 0.43 | 0.38 | 0.33 |
|  | 25 – 39 | 1.00 | 0.94 | 0.87 | 0.80 | 0.73 | 0.67 | 0.61 | 0.56 | 0.51 | 0.46 | 0.42 |
|  | 40 – 64 | 1.00 | 0.96 | 0.92 | 0.87 | 0.81 | 0.75 | 0.70 | 0.66 | 0.62 | 0.58 | 0.55 |
|  | 65 – 119 | 1.00 | 0.98 | 0.95 | 0.91 | 0.87 | 0.82 | 0.78 | 0.75 | 0.72 | 0.69 | 0.66 |
|  | 120 – 199 | 1.00 | 0.99 | 0.97 | 0.95 | 0.92 | 0.89 | 0.86 | 0.84 | 0.81 | 0.79 | 0.78 |
| 12–18 | 17 – 24 | 1.00 | 0.94 | 0.88 | 0.81 | 0.74 | 0.67 | 0.61 | 0.55 | 0.49 | 0.43 | 0.37 |
|  | 25 – 39 | 1.00 | 0.96 | 0.91 | 0.86 | 0.80 | 0.74 | 0.68 | 0.62 | 0.56 | 0.50 | 0.44 |
|  | 40 – 64 | 1.00 | 0.98 | 0.94 | 0.90 | 0.85 | 0.80 | 0.75 | 0.70 | 0.67 | 0.62 | 0.58 |
|  | 65 – 119 | 1.00 | 0.99 | 0.96 | 0.93 | 0.90 | 0.86 | 0.82 | 0.79 | 0.75 | 0.72 | 0.69 |
|  | 120 – 199 | 1.00 | 1.00 | 0.98 | 0.96 | 0.94 | 0.92 | 0.89 | 0.86 | 0.83 | 0.80 | 0.78 |
|  | 200 or more | 1.00 | 1.00 | 1.00 | 0.98 | 0.97 | 0.95 | 0.93 | 0.91 | 0.90 | 0.88 | 0.87 |
| 18–24 | 40 – 64 | 1.00 | 1.00 | 0.96 | 0.93 | 0.89 | 0.84 | 0.79 | 0.74 | 0.69 | 0.64 | 0.59 |
|  | 65 – 119 | 1.00 | 1.00 | 0.97 | 0.94 | 0.92 | 0.89 | 0.86 | 0.83 | 0.80 | 0.76 | 0.73 |
|  | 120 – 199 | 1.00 | 1.00 | 0.99 | 0.98 | 0.96 | 0.94 | 0.92 | 0.90 | 0.88 | 0.86 | 0.85 |
|  | 200 or more | 1.00 | 1.00 | 1.00 | 1.00 | 0.98 | 0.96 | 0.95 | 0.93 | 0.92 | 0.92 | 0.91 |
| 24–30 | 40 – 64 | 1.00 | 0.98 | 0.94 | 0.90 | 0.85 | 0.80 | 0.74 | 0.69 | 0.65 | 0.61 | 0.58 |
|  | 65 – 119 | 1.00 | 0.99 | 0.97 | 0.93 | 0.90 | 0.86 | 0.82 | 0.79 | 0.76 | 0.73 | 0.71 |
|  | 120 – 199 | 1.00 | 1.00 | 0.98 | 0.96 | 0.94 | 0.92 | 0.89 | 0.87 | 0.85 | 0.83 | 0.81 |
|  | 200 or more | 1.00 | 1.00 | 0.99 | 0.98 | 0.97 | 0.95 | 0.93 | 0.92 | 0.90 | 0.89 | 0.89 |
| 30–35 | 40 – 64 | 1.00 | 0.96 | 0.92 | 0.86 | 0.80 | 0.74 | 0.68 | 0.64 | 0.60 | 0.57 | 0.55 |
|  | 65 – 119 | 1.00 | 0.98 | 0.95 | 0.90 | 0.86 | 0.81 | 0.76 | 0.72 | 0.68 | 0.65 | 0.62 |
|  | 120 – 199 | 1.00 | 0.99 | 0.97 | 0.95 | 0.92 | 0.88 | 0.85 | 0.82 | 0.80 | 0.78 | 0.77 |
|  | 200 or more | 1.00 | 1.00 | 0.98 | 0.97 | 0.95 | 0.93 | 0.90 | 0.89 | 0.87 | 0.86 | 0.85 |
| 35–42 | 40 – 64 | 1.00 | 0.95 | 0.89 | 0.82 | 0.75 | 0.69 | 0.63 | 0.58 | 0.53 | 0.49 | 0.46 |
|  | 65 – 119 | 1.00 | 0.97 | 0.93 | 0.88 | 0.82 | 0.77 | 0.71 | 0.67 | 0.63 | 0.59 | 0.56 |
|  | 120 – 199 | 1.00 | 0.98 | 0.96 | 0.93 | 0.89 | 0.85 | 0.81 | 0.78 | 0.76 | 0.73 | 0.71 |
|  | 200 or more | 1.00 | 0.99 | 0.98 | 0.96 | 0.93 | 0.90 | 0.87 | 0.84 | 0.82 | 0.80 | 0.78 |

Notes: 1. $A_1$ = Cross-sectional area of main member before boring or grooving.
2. $A_2$ = Sum of cross-sectional areas of metal side plates before drilling.

By permission of NFPA

## Condition of Use

The condition of use factors for bolts, lags, shear plates, and split rings are similar to those used in nailed connections, with one very notable exception. As timber seasons in place, it shrinks across the grain. If the member is restrained from doing so by a metal side plate, its only option is to split at the fasteners. This drastically reduces the fastener capacity, resulting in 60% reduction in design capacity. The situation is corrected by using separate splice plates with each row of fasteners across the grain. This is only one example of how tension perpendicular to the grain can influence the behavior of a connection which might appear to have no applied loads in a direction perpendicular to the grain.

# TIMBER DESIGN

## Table 9
### Special Condition of Use Factors for Connections Seasoning in Place

Factors apply when wood is at or above the fiber saturation point (wet) at time of fabrication but dries to a moisture content of 19 percent or less (dry) before full design load is applied. For wood partially seasoned when fabricated, adjusted intermediate values may be used.

| Arrangement of bolts or lag screws | Type of splice plate | Modification factor |
|---|---|---|
| —One fastener only, or<br>—Two or more fasteners placed in a single line parallel to grain, or<br>—Fasteners placed in two or more lines parallel to grain with separate splice plates for each line | Wood or metal | 1.0 |
| —All other arrangements | Wood or metal | 0.4 |

By permission of NFPA

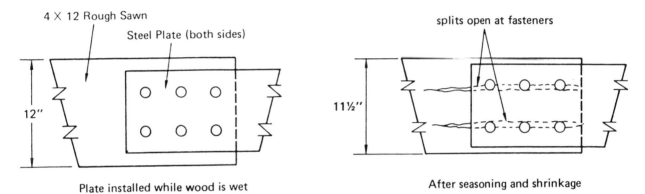

**Figure 11
Seasoning in Place**

## Net Section

At connections, member stress levels have to be based on a cross sectional area with the bolt or lag holes removed. Assume bolt holes have a 1/16" larger diameter than the bolt. Subtract the diameter times the penetration for lag screws. Bolts or lags in adjacent rows of staggered bolts or lags are considered to be at the same section unless they are spaced more than eight diameters apart in the rows.

Bolts or lag screws in adjacent staggered rows are considered to be at the same section if they fall within a piece of the member which is 4 diameters long.

## Angle of Load to Grain

Lag screw and bolt design capacities are tabulated for use in parallel and perpendicular to the grain directions. For other load directions, use the Hankinson formula to interpolate.

### E. Lag Screws

A lag screw will only develop its full lateral and withdrawal capacities if installed correctly. The proper installation tool is a wrench, not a hammmer. A pilot hole must be drilled and sized according to precise standards. Lag bolt capacities are subject to the condition of use factors found in table 4.

#### Withdrawal Loads

The withdrawal capacity of a lag screw is a function of its diameter, the holding member's specific gravity, and the effective threaded length in the holding member. The dimension "T-E", found in appendix 3, is the length of effective thread. This effective length, multiplied by the allowable load per inch found in table 10, is the allowable withdrawal load for a lag screw.

It is possible for the lag screw-to-timber capacity to exceed that of the lag screw itself. If the effective threaded length is less than 7 diameters in Group 1 species, 8 diameters in Group 2, 10 diameters in Group 3, and 11 diameters in Group 4, the lag screw should not break before it is pulled out.

- **Withdrawal from End Grain:** If a lag screw must unavoidably be loaded in withdrawal while installed in end grain, the capacity should be reduced by 25%.

#### Lateral Loads in Lag Screws

Tables 11 and 12 give the allowable lateral loads, parallel and perpendicular to the grain, for all lag screw diameters and lengths in the four species groups found in table 3. The only difference in the tables is the side member material. If the load is at an angle to the grain between the two listed values, the Hankinson formula should be used.

- **End Grain Installation:** The listed values are for side grain installation. If the lag screw is in the end grain, reduce the load capacity by a third.

- Combined Loadings: Just as with nails, no interaction between combined lateral and withdrawal loads needs to be considered. Each load component is compared with its own respective design value.

- Lag Screw Spacing: The design values of tables 11 and 12 assume the lag screws are spaced according to the spacing requirements for bolts.

### Table 10
### Allowable Withdrawal Loads -- Lag Screws

Normal load duration, dry service conditions

Design values for load in withdrawal in pounds per inch of penetration of threaded part into side grain of member holding point.

$D$ = the shank diameter in inches.
$G$ = specific gravity of the wood based on weight and volume when oven-dry.

| Specific gravity $G$ | Lag screw diameter $D$ | | | | | | | | | | | |
|---|---|---|---|---|---|---|---|---|---|---|---|---|
| | 1/4 | 5/16 | 3/8 | 7/16 | 1/2 | 9/16 | 5/8 | 3/4 | 7/8 | 1 | 1-1/8 | 1-1/4 |
| | 0.250 | 0.3125 | 0.375 | 0.4375 | 0.500 | 0.5625 | 0.625 | 0.750 | 0.875 | 1.000 | 1.125 | 1.250 |
| 0.75 | 413 | 489 | 560 | 629 | 695 | 759 | 822 | 942 | 1058 | 1169 | 1277 | 1382 |
| 0.68 | 357 | 422 | 484 | 543 | 600 | 656 | 709 | 813 | 913 | 1009 | 1103 | 1193 |
| 0.67 | 349 | 413 | 473 | 531 | 587 | 641 | 694 | 796 | 893 | 987 | 1078 | 1167 |
| 0.66 | 341 | 403 | 463 | 519 | 574 | 627 | 678 | 778 | 873 | 965 | 1054 | 1141 |
| 0.62 | 311 | 367 | 421 | 473 | 523 | 571 | 618 | 708 | 795 | 879 | 960 | 1039 |
| 0.55 | 260 | 307 | 352 | 395 | 437 | 477 | 516 | 592 | 664 | 734 | 802 | 868 |
| 0.54 | 253 | 299 | 342 | 384 | 425 | 464 | 502 | 576 | 646 | 714 | 780 | 844 |
| 0.51 | 232 | 274 | 314 | 353 | 390 | 426 | 461 | 528 | 593 | 656 | 716 | 775 |
| 0.49 | 218 | 258 | 296 | 332 | 367 | 401 | 434 | 498 | 559 | 617 | 674 | 730 |
| 0.48 | 212 | 250 | 287 | 322 | 356 | 389 | 421 | 482 | 542 | 599 | 654 | 708 |
| 0.47 | 205 | 242 | 278 | 312 | 345 | 377 | 408 | 467 | 525 | 580 | 634 | 686 |
| 0.46 | 199 | 235 | 269 | 302 | 334 | 365 | 395 | 453 | 508 | 562 | 613 | 664 |
| 0.45 | 192 | 227 | 260 | 292 | 323 | 353 | 382 | 438 | 492 | 543 | 594 | 642 |
| 0.44 | 186 | 220 | 252 | 283 | 312 | 341 | 369 | 423 | 475 | 525 | 574 | 621 |
| 0.43 | 179 | 212 | 243 | 273 | 302 | 330 | 357 | 409 | 459 | 508 | 554 | 600 |
| 0.42 | 173 | 205 | 235 | 264 | 291 | 318 | 344 | 395 | 443 | 490 | 535 | 579 |
| 0.41 | 167 | 198 | 226 | 254 | 281 | 307 | 332 | 381 | 428 | 473 | 516 | 559 |
| 0.40 | 161 | 190 | 218 | 245 | 271 | 296 | 320 | 367 | 412 | 455 | 497 | 538 |
| 0.39 | 155 | 183 | 210 | 236 | 261 | 285 | 308 | 353 | 397 | 438 | 479 | 518 |
| 0.38 | 149 | 176 | 202 | 227 | 251 | 274 | 296 | 340 | 381 | 422 | 461 | 498 |
| 0.37 | 143 | 169 | 194 | 218 | 241 | 263 | 285 | 326 | 367 | 405 | 443 | 479 |
| 0.36 | 137 | 163 | 186 | 209 | 231 | 253 | 273 | 313 | 352 | 389 | 425 | 460 |
| 0.35 | 132 | 156 | 179 | 200 | 222 | 242 | 262 | 300 | 337 | 373 | 407 | 441 |
| 0.33 | 121 | 143 | 164 | 184 | 203 | 222 | 240 | 275 | 309 | 341 | 373 | 403 |
| 0.31 | 110 | 130 | 149 | 167 | 185 | 202 | 218 | 250 | 281 | 311 | 339 | 367 |

By permission of NFPA

## Table 11
### Allowable Lateral Loads in Lag Screws  Wood Side Pieces

Normal load duration, dry service conditions

| Thickness of side member (inches) | Length of lag screw (inches) | Diameter of lag screw shank (inches) | Species Group (See Table 3) | | | | | | | |
|---|---|---|---|---|---|---|---|---|---|---|
| | | | GROUP I | | GROUP II | | GROUP III | | GROUP IV | |
| | | | Total lateral load per lag screw in single shear (pounds) | | Total lateral load per lag screw in single shear (pounds) | | Total lateral load per lag screw in single shear (pounds) | | Total lateral load per lag screw in single shear (pounds) | |
| | | | Parallel to grain | Perpendicular to grain | Parallel to grain | Perpendicular to grain | Parallel to grain | Perpendicular to grain | Parallel to grain | Perpendicular to grain |
| 1½" | 4" | 1/4 | 200 | 200 | 170 | 170 | 130 | 130 | 100 | 100 |
| | | 5/16 | 290 | 240 | 220 | 180 | 150 | 130 | 120 | 110 |
| | | 3/8 | 330 | 250 | 250 | 190 | 180 | 140 | 140 | 110 |
| | | 7/16 | 370 | 260 | 280 | 190 | 200 | 140 | 160 | 110 |
| | | 1/2 | 390 | 250 | 290 | 190 | 210 | 140 | 170 | 110 |
| | | 5/8 | 470 | 280 | 360 | 210 | 260 | 160 | 200 | 120 |
| | 5" | 1/4 | 240 | 230 | 200 | 200 | 180 | 180 | 160 | 160 |
| | | 5/16 | 340 | 290 | 290 | 250 | 240 | 200 | 190 | 160 |
| | | 3/8 | 440 | 340 | 380 | 290 | 270 | 210 | 220 | 170 |
| | | 7/16 | 550 | 380 | 420 | 290 | 300 | 210 | 240 | 170 |
| | | 1/2 | 580 | 380 | 440 | 280 | 310 | 200 | 250 | 160 |
| | | 5/8 | 710 | 420 | 530 | 320 | 380 | 230 | 310 | 180 |
| | 6" | 1/4 | 270 | 260 | 230 | 220 | 210 | 200 | 180 | 180 |
| | | 5/16 | 380 | 320 | 330 | 280 | 290 | 250 | 260 | 220 |
| | | 3/8 | 490 | 370 | 420 | 320 | 370 | 280 | 300 | 230 |
| | | 7/16 | 600 | 420 | 520 | 360 | 410 | 280 | 330 | 230 |
| | | 1/2 | 700 | 460 | 600 | 390 | 430 | 280 | 340 | 220 |
| | | 5/8 | 850 | 510 | 710 | 430 | 510 | 310 | 410 | 250 |
| | 7" | 1/4 | 280 | 270 | 240 | 230 | 210 | 210 | 190 | 180 |
| | | 5/16 | 400 | 340 | 350 | 300 | 310 | 270 | 280 | 230 |
| | | 3/8 | 530 | 400 | 460 | 350 | 410 | 310 | 360 | 270 |
| | | 7/16 | 650 | 450 | 560 | 390 | 500 | 350 | 420 | 300 |
| | | 1/2 | 760 | 490 | 660 | 430 | 550 | 360 | 440 | 290 |
| | | 5/8 | 910 | 540 | 780 | 470 | 640 | 380 | 510 | 310 |
| 2½" | 6" | 3/8 | 450 | 340 | 380 | 290 | 270 | 210 | 220 | 170 |
| | | 7/16 | 590 | 410 | 440 | 310 | 320 | 220 | 250 | 180 |
| | | 1/2 | 620 | 410 | 470 | 310 | 340 | 220 | 270 | 180 |
| | | 5/8 | 730 | 440 | 550 | 330 | 390 | 240 | 320 | 190 |
| | | 3/4 | 830 | 460 | 630 | 350 | 450 | 250 | 360 | 200 |
| | | 7/8 | 950 | 490 | 720 | 370 | 510 | 270 | 410 | 210 |
| | | 1 | 1060 | 530 | 800 | 400 | 570 | 290 | 460 | 230 |
| | 7" | 3/8 | 500 | 380 | 430 | 330 | 380 | 290 | 300 | 230 |
| | | 7/16 | 670 | 470 | 580 | 410 | 430 | 300 | 350 | 240 |
| | | 1/2 | 830 | 540 | 650 | 420 | 460 | 300 | 370 | 240 |
| | | 5/8 | 1000 | 600 | 750 | 450 | 540 | 320 | 430 | 260 |
| | | 3/4 | 1120 | 620 | 850 | 470 | 610 | 330 | 490 | 270 |
| | | 7/8 | 1280 | 660 | 970 | 500 | 690 | 360 | 550 | 290 |
| | | 1 | 1440 | 720 | 1090 | 540 | 780 | 390 | 620 | 310 |
| | 8" | 3/8 | 560 | 420 | 480 | 370 | 430 | 330 | 380 | 290 |
| | | 7/16 | 730 | 510 | 630 | 440 | 560 | 390 | 450 | 320 |
| | | 1/2 | 890 | 580 | 770 | 500 | 600 | 390 | 480 | 310 |
| | | 5/8 | 1230 | 740 | 970 | 580 | 700 | 420 | 560 | 340 |
| | | 3/4 | 1440 | 790 | 1090 | 600 | 780 | 430 | 630 | 340 |
| | | 7/8 | 1610 | 840 | 1220 | 630 | 870 | 450 | 700 | 360 |
| | | 1 | 1810 | 910 | 1370 | 690 | 980 | 490 | 790 | 390 |
| | 9" | 3/8 | 600 | 460 | 520 | 400 | 470 | 350 | 410 | 310 |
| | | 7/16 | 790 | 550 | 680 | 480 | 610 | 430 | 540 | 380 |
| | | 1/2 | 960 | 630 | 830 | 540 | 740 | 480 | 600 | 390 |
| | | 5/8 | 1310 | 790 | 1130 | 680 | 860 | 520 | 690 | 420 |
| | | 3/4 | 1680 | 920 | 1350 | 740 | 960 | 530 | 770 | 430 |
| | | 7/8 | 1940 | 1010 | 1470 | 760 | 1050 | 550 | 840 | 440 |
| | | 1 | 2190 | 2000 | 1660 | 830 | 1190 | 590 | 950 | 480 |

# Table 12
## Allowable Lateral Loads in Lag Screws Metal Side Pieces

Normal load duration, dry service conditions

| Length of lag screw (inches) | Diameter of lag screw shank (inches) | GROUP I Parallel to grain | GROUP I Perpendicular to grain | GROUP II Parallel to grain | GROUP II Perpendicular to grain | GROUP III Parallel to grain | GROUP III Perpendicular to grain | GROUP IV Parallel to grain | GROUP IV Perpendicular to grain |
|---|---|---|---|---|---|---|---|---|---|
| 3" | 1/4 | 240 | 190 | 210 | 160 | 160 | 120 | 130 | 100 |
|  | 5/16 | 350 | 240 | 270 | 180 | 190 | 130 | 150 | 100 |
|  | 3/8 | 420 | 250 | 320 | 190 | 230 | 140 | 180 | 110 |
|  | 7/16 | 480 | 270 | 360 | 200 | 260 | 140 | 210 | 120 |
|  | 1/2 | 540 | 280 | 400 | 210 | 290 | 150 | 230 | 120 |
|  | 5/8 | 650 | 310 | 490 | 230 | 350 | 170 | 280 | 130 |
| 4" | 1/4* | 270 | 210 | 240 | 180 | 210 | 160 | 190 | 150 |
|  | 5/16 | 410 | 280 | 350 | 240 | 290 | 200 | 230 | 160 |
|  | 3/8 | 570 | 350 | 480 | 290 | 340 | 210 | 280 | 170 |
|  | 7/16 | 730 | 410 | 550 | 310 | 390 | 220 | 310 | 180 |
|  | 1/2 | 810 | 420 | 610 | 320 | 440 | 230 | 350 | 180 |
|  | 5/8 | 980 | 470 | 740 | 360 | 530 | 250 | 430 | 200 |
| 5" | 5/16 | 440 | 300 | 380 | 260 | 340 | 230 | 300 | 200 |
|  | 3/8 | 620 | 380 | 530 | 320 | 470 | 290 | 380 | 230 |
|  | 7/16 | 810 | 460 | 700 | 390 | 530 | 300 | 430 | 240 |
|  | 1/2 | 1040 | 540 | 840 | 440 | 600 | 310 | 480 | 250 |
|  | 5/8 | 1330 | 640 | 1010 | 480 | 720 | 350 | 580 | 280 |
|  | 3/4 | 1550 | 680 | 1170 | 520 | 840 | 370 | 670 | 300 |
| 6" | 5/16* | 450 | 300 | 390 | 260 | 340 | 230 | 300 | 210 |
|  | 3/8 | 630 | 390 | 550 | 330 | 490 | 300 | 430 | 260 |
|  | 7/16 | 850 | 480 | 730 | 410 | 660 | 370 | 540 | 300 |
|  | 1/2 | 1100 | 570 | 950 | 490 | 760 | 400 | 610 | 320 |
|  | 5/8 | 1640 | 790 | 1290 | 620 | 920 | 440 | 740 | 350 |
|  | 3/4 | 1990 | 870 | 1500 | 660 | 1070 | 470 | 860 | 380 |
| 7" | 3/8* | 640 | 390 | 560 | 340 | 500 | 300 | 440 | 270 |
|  | 7/16 | 870 | 490 | 750 | 420 | 670 | 380 | 590 | 330 |
|  | 1/2 | 1120 | 580 | 970 | 500 | 870 | 450 | 740 | 380 |
|  | 5/8 | 1710 | 820 | 1480 | 710 | 1130 | 540 | 900 | 430 |
|  | 3/4 | 2380 | 1050 | 1840 | 810 | 1310 | 580 | 1050 | 460 |
| 8" | 7/16* | 880 | 490 | 760 | 420 | 680 | 380 | 600 | 330 |
|  | 1/2 | 1140 | 590 | 980 | 510 | 880 | 460 | 780 | 400 |
|  | 5/8 | 1750 | 840 | 1510 | 720 | 1320 | 630 | 1060 | 510 |
|  | 3/4 | 2470 | 1090 | 2130 | 940 | 1560 | 690 | 1250 | 550 |
|  | 7/8 | 3260 | 1360 | 2480 | 1030 | 1770 | 740 | 1420 | 590 |
| 9" | 1/2* | 1150 | 600 | 990 | 510 | 890 | 460 | 780 | 410 |
|  | 5/8 | 1770 | 850 | 1530 | 730 | 1370 | 660 | 1210 | 580 |
|  | 3/4 | 2510 | 1100 | 2160 | 950 | 1790 | 790 | 1440 | 630 |
|  | 7/8 | 3360 | 1400 | 2880 | 1200 | 2060 | 860 | 1650 | 690 |
| 10" | 5/8* | 1790 | 860 | 1550 | 740 | 1380 | 660 | 1220 | 590 |
|  | 3/4 | 2550 | 1120 | 2200 | 970 | 1970 | 870 | 1630 | 720 |
|  | 7/8 | 3430 | 1420 | 2960 | 1230 | 2340 | 970 | 1880 | 780 |
|  | 1 | 4410 | 1770 | 3680 | 1470 | 2640 | 1050 | 2110 | 850 |
| 11" | 3/4* | 2570 | 1130 | 2220 | 980 | 1990 | 880 | 1750 | 770 |
|  | 7/8 | 3470 | 1440 | 3000 | 1250 | 2620 | 1090 | 2100 | 870 |
|  | 1 | 4490 | 1800 | 3880 | 1550 | 2960 | 1180 | 2370 | 950 |
| 12" | 7/8 | 3490 | 1450 | 3020 | 1260 | 2700 | 1120 | 2320 | 960 |
|  | 1 | 4520 | 1810 | 3900 | 1560 | 3260 | 1310 | 2620 | 1050 |
|  | 1-1/8 | 5670 | 2270 | 4890 | 1960 | 3630 | 1450 | 2910 | 1170 |
| 13" | 7/8* | 3510 | 1460 | 3030 | 1260 | 2710 | 1130 | 2390 | 1000 |
|  | 1 | 4550 | 1820 | 3930 | 1570 | 3520 | 1410 | 2870 | 1150 |
|  | 1-1/8 | 5710 | 2280 | 4930 | 1970 | 3980 | 1590 | 3200 | 1280 |
| 14" | 1 | 4570 | 1830 | 3950 | 1580 | 3530 | 1410 | 3110 | 1250 |
|  | 1-1/8 | 5750 | 2300 | 4960 | 1980 | 4330 | 1730 | 3470 | 1390 |
|  | 1-1/4 | 7030 | 2810 | 6070 | 2430 | 4750 | 1900 | 3810 | 1520 |
| 15" | 1 | 4590 | 1830 | 3960 | 1580 | 3540 | 1420 | 3130 | 1250 |
|  | 1-1/8 | 5770 | 2310 | 4990 | 1990 | 4460 | 1780 | 3750 | 1500 |
|  | 1-1/4 | 7070 | 2830 | 6110 | 2440 | 5130 | 2050 | 4120 | 1650 |
| 16" | 1* | 4590 | 1830 | 3960 | 1580 | 3540 | 1420 | 3130 | 1250 |
|  | 1-1/8* | 5790 | 2320 | 5000 | 2000 | 4480 | 1790 | 3950 | 1580 |
|  | 1-1/4* | 7120 | 2850 | 6150 | 2460 | 5500 | 2200 | 4430 | 1770 |

*Greater lengths do not provide higher loads.

## Example 5

Select the 2 lag screws for attaching the light to the field house roof.

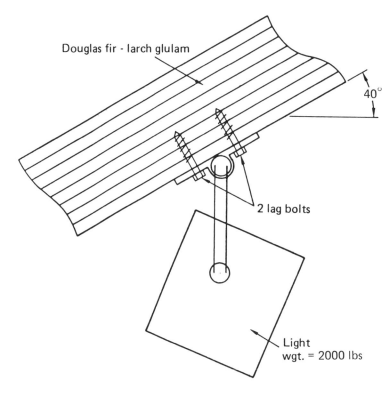

It is necessary to determine the lateral and withdrawal components of the applied load. Select a trial lag screw and compare its capacities with those required. It would be wasteful to use a lag screw which is at its design capacity in one component, while significantly underloaded in the other.

Table 3 has Douglas fir-larch in group II, with a specific gravity, G = .51. The CUF = 1.0, from table 4. Table 2 gives a LDF = .9 for dead loads.

The required lateral capacity per lag screw is

(2000) lb $(\sin 40°)/(.9)$ (2) lag screws = 714 lb/lag screw

The required withdrawal capacity per lag screw is

(2000) lb $(\cos 40°)/(.9)$ (2) lag screws = 851 lb/lag screw

Try a 6"x7/16" lag screw. The lateral capacity = 730 lb, parallel to grain, (table 12). The withdrawal capacity = (T-E) X allowable lb/in. The effective penetration, T-E = 3-7/32" (appendix 3). The allowable withdrawal loading = 353 lb/in (table 10). The withdrawal capacity is therefore equal to (3.219) in (353) lb/in = 1136 lb.

The 6"x7/16" lag screw satisfies both the applied withdrawal and lateral loads.

## Example 6

The lag screw shown is intended to hold the wharf against wind load. What is its capacity?

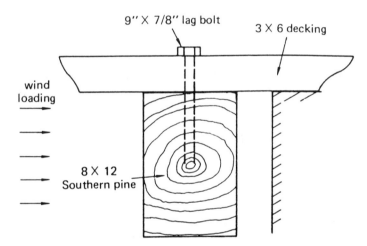

Determine the lag screw capacity and modify for condition of use and load duration.

Southern pine is a group II species (table 3). The wind load duration factor is 1.33 (table 2). Assuming the wharf is subjected to a wet-dry cycle, table 4 gives a CUF = .75. The lag screw capacity is 760 lb, found in table 11.

Lag screw capacity = (760) lb (1.33) (.75) = 760 lb

### F. Bolts

The shear capacity of a bolt depends on the wood species involved, the length and diameter of the bolt, the load directions, and the number of members and their configuration. Table 13 is a straightforward list of design values based on an assumed layout for a typical connection. The listed values are for dry conditions and normal load duration. These assumptions are readily modified to those of the design situation, using the condition of use and load duration factors.

## Reading Load Tables

An assumption inherent to table 13 is that of two side members, each half as thick as the main member sandwiched between them. This "double-shear" assumption is not true for connections with only two members, or with more than three members. Other connection configurations are dealt with by using a modified or substitute main member thickness in table 13, while still using the actual bolt size and member species.

In outline form, this is the length of bolt in main member, l, with which to enter table 13.

A. All members aligned with each other

1. Double shear

$t_{\text{side members}} = (.5)(t_{\text{main member}})$
    Table assumption, use tabulated value

$t_{\text{side members}} > (.5)(t_{\text{main member}})$
    No increase, use tabulated value

$t_{\text{side members}} < (.5)(t_{\text{main member}})$
    Use tabulated value with $t = (2)(t_{\text{thinnest side member}})$

2. Single shear (only two members)

    Equal thickness
        Use $(.5)$(tabulated value), with $t = t_{\text{member}}$

    Unequal thickness
        Use **lesser** of:
            $(.5)$(tabulated value), with $t = t_{\text{thicker member}}$
            $(.5)$(tabulated value), with $t = (2)(t_{\text{thinner member}})$

# TIMBER DESIGN

B. Members not aligned with each other

> Use **lesser** of:
> > tabulated value for main member
> > tabulated value, with $t = (2)(t_{side\ member})$ and loaded in the appropriate direction. Use the Hankinson formula.

> NOTE: Connection details are extremely important in timber, because of the effect they can have on load direction. See figures 12 through 15 for an illustration.

C. Members of different material

1. Different species of timber at same connection
   Use lesser of:
   - tabulated value: all members of side member species
   - tabulated value: all members of main member species

2. Metal members in connection

   Metal side member(s) (straps or plates)

   > Parallel to grain: increase tabulated value for main members by:
   > a) 75% for bolts ½ inch or less in diameter
   > b) 25% for bolts 1½ inch or larger in diameter
   > c) proportional percentage for intermediate diameter bolts (i.e. between ½" and 1½")
   >
   > Perpendicular to grain: no increase in tabulated values

   Metal main member(s)

   > Parallel to grain: increase the tabulated value for a piece with $t = (2)(t_{side\ member})$ by:
   > a) 75% for bolts ½ inch or less in diameter
   > b) 25% for bolts 1½ inch or larger in diameter
   > c) proportional percentage for intermediate diameter bolts
   >
   > Perpendicular to grain: no increase in tabulated values with $t = (2)(t_{side\ member})$

[NOTE: The 75% increase in allowable loads with steel side members is as much as a three-fold increase over the former 25% value. This is a controversial issue, and should be checked carefully against the most recent codes.]

## Table 13  Lateral Loads in Bolts -- Design Values

Design values, in pounds, on one bolt loaded at both ends (double shear)* for following species

| | | | | 1 DOUGLAS FIR-LARCH (Dense), SOUTHERN PINE (Dense) | | 2 ASH, Commercial White, HICKORY | | 3 CALIFORNIA REDWOOD (Close grain), DOUGLAS FIR-LARCH, SOUTHERN PINE, SOUTHERN CYPRESS | | 4 BEECH, BIRCH, Sweet & Yellow, MAPLE, Black & Sugar | | 5 OAK, Red & White | | 6 DOUGLAS FIR SOUTH | |
|---|---|---|---|---|---|---|---|---|---|---|---|---|---|---|---|
| Length of bolt in main member $\ell$ | Diameter of bolt D | $\ell/D$ | Projected area of bolt $A=\ell \times D$ | Parallel to grain P | Perpendicular to grain Q | Parallel to grain P | Perpendicular to grain Q | Parallel to grain P | Perpendicular to grain Q | Parallel to grain P | Perpendicular to grain Q | Parallel to grain P | Perpendicular to grain Q | Parallel to grain P | Perpendicular to grain Q |
| 1-1/2 | 1/2 | 3.00 | .750 | 1100 | 500 | 1080 | 780 | 940 | 430 | 900 | 480 | 830 | 650 | 870 | 370 |
| | 5/8 | 2.40 | .938 | 1380 | 570 | 1360 | 880 | 1180 | 490 | 1130 | 540 | 1050 | 730 | 1090 | 420 |
| | 3/4 | 2.00 | 1.125 | 1660 | 630 | 1630 | 980 | 1420 | 540 | 1360 | 600 | 1260 | 820 | 1310 | 470 |
| | 7/8 | 1.71 | 1.313 | 1940 | 700 | 1910 | 1080 | 1660 | 600 | 1590 | 670 | 1470 | 900 | 1530 | 520 |
| | 1 | 1.50 | 1.500 | 2220 | 760 | 2180 | 1170 | 1890 | 650 | 1820 | 730 | 1690 | 980 | 1750 | 570 |
| 2 | 1/2 | 4.00 | 1.000 | 1370 | 670 | 1340 | 1030 | 1170 | 570 | 1120 | 640 | 1040 | 870 | 1140 | 500 |
| | 5/8 | 3.20 | 1.250 | 1810 | 760 | 1780 | 1170 | 1550 | 650 | 1480 | 720 | 1380 | 980 | 1450 | 560 |
| | 3/4 | 2.67 | 1.500 | 2200 | 840 | 2170 | 1300 | 1880 | 720 | 1810 | 810 | 1680 | 1090 | 1750 | 630 |
| | 7/8 | 2.29 | 1.750 | 2580 | 930 | 2540 | 1430 | 2210 | 790 | 2120 | 890 | 1960 | 1200 | 2040 | 690 |
| | 1 | 2.00 | 2.000 | 2960 | 1010 | 2910 | 1560 | 2520 | 870 | 2420 | 970 | 2250 | 1310 | 2330 | 750 |
| 2-1/2 | 1/2 | 5.00 | 1.250 | 1480 | 830 | 1450 | 1280 | 1260 | 720 | 1210 | 800 | 1120 | 1070 | 1290 | 620 |
| | 5/8 | 4.00 | 1.563 | 2140 | 950 | 2100 | 1460 | 1820 | 810 | 1750 | 900 | 1620 | 1220 | 1780 | 710 |
| | 3/4 | 3.33 | 1.875 | 2700 | 1050 | 2650 | 1630 | 2310 | 900 | 2210 | 1010 | 2050 | 1360 | 2180 | 790 |
| | 7/8 | 2.86 | 2.188 | 3210 | 1160 | 3150 | 1790 | 2740 | 990 | 2630 | 1110 | 2440 | 1500 | 2550 | 860 |
| | 1 | 2.50 | 2.500 | 3680 | 1270 | 3620 | 1960 | 3150 | 1080 | 3020 | 1210 | 2800 | 1640 | 2920 | 940 |
| 3 | 1/2 | 6.00 | 1.500 | 1490 | 970 | 1460 | 1460 | 1270 | 860 | 1220 | 960 | 1130 | 1130 | 1330 | 750 |
| | 5/8 | 4.80 | 1.875 | 2290 | 1130 | 2250 | 1750 | 1960 | 970 | 1880 | 1090 | 1740 | 1460 | 1980 | 850 |
| | 3/4 | 4.00 | 2.250 | 3080 | 1270 | 3020 | 1950 | 2630 | 1080 | 2520 | 1210 | 2340 | 1630 | 2560 | 940 |
| | 7/8 | 3.43 | 2.625 | 3760 | 1390 | 3700 | 2150 | 3220 | 1190 | 3080 | 1330 | 2860 | 1800 | 3040 | 1040 |
| | 1 | 3.00 | 3.000 | 4390 | 1520 | 4310 | 2350 | 3750 | 1300 | 3600 | 1450 | 3340 | 1960 | 3500 | 1130 |
| 3-1/2 | 1/2 | 7.00 | 1.750 | 1490 | 1120 | 1460 | 1460 | 1270 | 980 | 1220 | 1090 | 1130 | 1130 | 1330 | 850 |
| | 5/8 | 5.60 | 2.188 | 2320 | 1310 | 2280 | 2020 | 1980 | 1130 | 1900 | 1270 | 1770 | 1690 | 2060 | 990 |
| | 3/4 | 4.67 | 2.625 | 3280 | 1470 | 3220 | 2270 | 2800 | 1260 | 2690 | 1410 | 2490 | 1900 | 2820 | 1100 |
| | 7/8 | 4.00 | 3.063 | 4190 | 1630 | 4120 | 2510 | 3580 | 1390 | 3430 | 1550 | 3180 | 2100 | 3480 | 1210 |
| | 1 | 3.50 | 3.500 | 5000 | 1770 | 4920 | 2740 | 4270 | 1520 | 4100 | 1690 | 3800 | 2290 | 4050 | 1320 |
| 4 | 1/2 | 8.00 | 2.000 | 1490 | 1010 | 1460 | 1460 | 1270 | 1010 | 1220 | 1130 | 1130 | 1130 | 1330 | 880 |
| | 5/8 | 6.40 | 2.500 | 2330 | 1410 | 2290 | 2180 | 1990 | 1290 | 1910 | 1440 | 1770 | 1770 | 2080 | 1120 |
| | 3/4 | 5.33 | 3.000 | 3340 | 1690 | 3280 | 2600 | 2850 | 1440 | 2740 | 1610 | 2540 | 2180 | 2950 | 1260 |
| | 7/8 | 4.57 | 3.500 | 4440 | 1850 | 4360 | 2860 | 3790 | 1590 | 3640 | 1770 | 3370 | 2390 | 3800 | 1380 |
| | 1 | 4.00 | 4.000 | 5470 | 2030 | 5380 | 3130 | 4670 | 1730 | 4480 | 1940 | 4160 | 2620 | 4540 | 1510 |
| 4-1/2 | 5/8 | 7.20 | 2.813 | 2330 | 1440 | 2290 | 2230 | 1990 | 1400 | 1910 | 1560 | 1770 | 1770 | 2070 | 1220 |
| | 3/4 | 6.00 | 3.375 | 3350 | 1830 | 3290 | 2820 | 2860 | 1620 | 2750 | 1810 | 2550 | 2360 | 2980 | 1410 |
| | 7/8 | 5.14 | 3.938 | 4540 | 2110 | 4460 | 3260 | 3880 | 1790 | 3720 | 2000 | 3450 | 2720 | 3980 | 1560 |
| | 1 | 4.50 | 4.500 | 5770 | 2280 | 5670 | 3520 | 4930 | 1950 | 4730 | 2180 | 4390 | 2940 | 4920 | 1700 |
| | 1-1/4 | 3.60 | 5.625 | 7970 | 2670 | 7830 | 4120 | 6810 | 2280 | 6530 | 2550 | 6060 | 3450 | 6490 | 1990 |
| 5-1/2 | 5/8 | 8.80 | 3.438 | 2330 | 1890 | 2290 | 2150 | 1990 | 1410 | 1910 | 1570 | 1770 | 1770 | 3080 | 1220 |
| | 3/4 | 7.33 | 4.125 | 3350 | 1930 | 3300 | 2980 | 2860 | 1880 | 2750 | 2100 | 2550 | 2490 | 2990 | 1630 |
| | 7/8 | 6.29 | 4.813 | 4570 | 2400 | 4490 | 3710 | 3900 | 2180 | 3740 | 2430 | 3470 | 3100 | 4070 | 1900 |
| | 1 | 5.50 | 5.500 | 5930 | 2760 | 5830 | 4260 | 5070 | 2380 | 4860 | 2660 | 4510 | 3560 | 5270 | 2080 |
| | 1-1/4 | 4.40 | 6.875 | 8930 | 3260 | 8780 | 5040 | 7630 | 2790 | 7320 | 3120 | 6790 | 4210 | 7580 | 2430 |
| 7-1/2 | 5/8 | 12.00 | 4.688 | 2330 | 1210 | 2290 | 1870 | 1990 | 1260 | 1910 | 1410 | 1770 | 1560 | 2070 | 1100 |
| | 3/4 | 10.00 | 5.625 | 3350 | 1930 | 3290 | 2710 | 2860 | 1820 | 2750 | 2030 | 2550 | 2260 | 2990 | 1580 |
| | 7/8 | 8.57 | 6.563 | 4560 | 2410 | 4480 | 3720 | 3900 | 2420 | 3740 | 2700 | 3470 | 3110 | 4060 | 2110 |
| | 1 | 7.50 | 7.500 | 5950 | 3090 | 5850 | 4760 | 5080 | 3030 | 4880 | 3390 | 4520 | 3980 | 5300 | 2640 |
| | 1-1/4 | 6.00 | 9.375 | 9310 | 4290 | 9150 | 6620 | 7950 | 3800 | 7630 | 4250 | 7080 | 5530 | 8290 | 3310 |
| 9-1/2 | 3/4 | 12.67 | 7.125 | 3350 | 1570 | 3290 | 2420 | 2860 | 1640 | 2740 | 1840 | 2540 | 2030 | 2990 | 1430 |
| | 7/8 | 10.86 | 8.313 | 4560 | 2180 | 4480 | 3360 | 3890 | 2270 | 3740 | 2540 | 3470 | 2810 | 4070 | 1980 |
| | 1 | 9.50 | 9.500 | 5950 | 2890 | 5850 | 4460 | 5080 | 2960 | 4880 | 3310 | 4520 | 3730 | 5310 | 2580 |
| | 1-1/4 | 7.60 | 11.875 | 9310 | 4510 | 9150 | 6960 | 7950 | 4450 | 7630 | 4970 | 7070 | 5820 | 8290 | 3870 |
| | 1-1/2 | 6.33 | 14.250 | 13420 | 6070 | 13190 | 9370 | 11470 | 5520 | 11000 | 6170 | 10200 | 7830 | 11960 | 4810 |
| 11-1/2 | 7/8 | 13.14 | 10.062 | 4560 | 1960 | 4490 | 3050 | 3900 | 2060 | 3750 | 2300 | 3470 | 2550 | 4070 | 1790 |
| | 1 | 11.50 | 11.500 | 5950 | 2650 | 5850 | 4080 | 5080 | 2770 | 4880 | 3090 | 4520 | 3410 | 5310 | 2410 |
| | 1-1/4 | 9.20 | 14.375 | 9310 | 4280 | 9160 | 6610 | 7960 | 4360 | 7630 | 4870 | 7080 | 5530 | 8300 | 3800 |
| | 1-1/2 | 7.67 | 17.250 | 13410 | 6210 | 13180 | 9590 | 11450 | 6140 | 10990 | 6860 | 10190 | 8010 | 11940 | 5350 |
| 13-1/2 | 1 | 13.50 | 13.500 | 5960 | 2390 | 5850 | 3730 | 5280 | 2530 | 4880 | 2830 | 4520 | 3120 | 5322 | 2190 |
| | 1-1/4 | 10.80 | 16.875 | 9300 | 3980 | 9150 | 6150 | 7950 | 4160 | 7630 | 4650 | 7070 | 5140 | 8300 | 3620 |
| | 1-1/2 | 9.00 | 20.250 | 13400 | 5950 | 13180 | 9190 | 11450 | 6040 | 10990 | 6750 | 10190 | 7680 | 11950 | 5260 |

*Three (3) member joint.

By permission of NFPA

# Timber Design

Design values, in pounds, on one bolt loaded at both ends (double shear)* for following species

| | | | | 7 SWEETGUM & TUPELO | | 8 EASTERN HEMLOCK-TAMARACK, CALIFORNIA REDWOOD (Open grain), HEM-FIR, WESTERN HEMLOCK | | 9 MOUNTAIN HEMLOCK, WESTERN CEDARS, NORTHERN PINE | | 10 SPRUCE-PINE-FIR, SITKA SPRUCE, YELLOW POPLAR, EASTERN SPRUCE, LODGEPOLE PINE | | 11 RED PINE, WESTERN WHITE PINE, PONDEROSA PINE-SUGAR PINE, EASTERN WHITE PINE, BALSAM FIR, IDAHO WHITE PINE | | 12 ASPEN, EASTERN COTTONWOOD, ENGELMANN SPRUCE-ALPINE FIR, NORTHERN WHITE CEDAR | |
|---|---|---|---|---|---|---|---|---|---|---|---|---|---|---|---|
| Length of bolt in main member $\ell$ | Diameter of bolt $D$ | $\ell/D$ | Projected area of bolt $A=\ell \times D$ | Parallel to grain $P$ | Perpendicular to grain $Q$ | Parallel to grain $P$ | Perpendicular to grain $Q$ | Parallel to grain $P$ | Perpendicular to grain $Q$ | Parallel to grain $P$ | Perpendicular to grain $Q$ | Parallel to grain $P$ | Perpendicular to grain $Q$ | Parallel to grain $P$ | Perpendicular to grain $Q$ |
| 1-1/2 | 1/2 | 3.00 | .750 | 810 | 410 | 800 | 270 | 750 | 300 | 680 | 280 | 630 | 190 | 530 | 210 |
| | 5/8 | 2.40 | .938 | 1010 | 460 | 1000 | 310 | 930 | 340 | 850 | 320 | 790 | 210 | 660 | 230 |
| | 3/4 | 2.00 | 1.125 | 1220 | 510 | 1200 | 350 | 1120 | 370 | 1020 | 350 | 950 | 240 | 800 | 260 |
| | 7/8 | 1.71 | 1.313 | 1420 | 560 | 1400 | 380 | 1310 | 410 | 1190 | 390 | 1110 | 260 | 930 | 290 |
| | 1 | 1.50 | 1.500 | 1620 | 620 | 1600 | 420 | 1490 | 450 | 1360 | 420 | 1260 | 290 | 1060 | 310 |
| 2 | 1/2 | 4.00 | 1.000 | 1050 | 540 | 1040 | 370 | 970 | 400 | 900 | 370 | 840 | 250 | 700 | 280 |
| | 5/8 | 3.20 | 1.250 | 1350 | 610 | 1330 | 410 | 1240 | 450 | 1130 | 420 | 1050 | 280 | 880 | 310 |
| | 3/4 | 2.67 | 1.500 | 1620 | 680 | 1600 | 460 | 1490 | 500 | 1360 | 470 | 1260 | 320 | 1060 | 350 |
| | 7/8 | 2.29 | 1.750 | 1890 | 750 | 1870 | 510 | 1740 | 550 | 1580 | 520 | 1480 | 350 | 1240 | 380 |
| | 1 | 2.00 | 2.000 | 2160 | 820 | 2130 | 550 | 1990 | 600 | 1810 | 560 | 1690 | 380 | 1420 | 420 |
| 2-1/2 | 1/2 | 5.00 | 1.250 | 1190 | 680 | 1180 | 460 | 1100 | 500 | 1080 | 470 | 1010 | 310 | 840 | 340 |
| | 5/8 | 4.00 | 1.563 | 1640 | 770 | 1620 | 520 | 1510 | 560 | 1410 | 530 | 1310 | 360 | 1100 | 390 |
| | 3/4 | 3.33 | 1.875 | 2020 | 850 | 1990 | 580 | 1860 | 620 | 1700 | 590 | 1580 | 400 | 1330 | 430 |
| | 7/8 | 2.86 | 2.188 | 2360 | 940 | 2330 | 630 | 2180 | 690 | 1980 | 650 | 1840 | 440 | 1550 | 480 |
| | 1 | 2.50 | 2.500 | 2700 | 1030 | 2670 | 690 | 2490 | 750 | 2260 | 700 | 2110 | 480 | 1770 | 520 |
| 3 | 1/2 | 6.00 | 1.500 | 1230 | 810 | 1210 | 550 | 1130 | 590 | 1160 | 560 | 1080 | 380 | 910 | 410 |
| | 5/8 | 4.80 | 1.875 | 1830 | 920 | 1810 | 620 | 1690 | 670 | 1640 | 630 | 1520 | 430 | 1280 | 470 |
| | 3/4 | 4.00 | 2.250 | 2370 | 1030 | 2340 | 690 | 2180 | 750 | 2030 | 700 | 1890 | 480 | 1590 | 520 |
| | 7/8 | 3.43 | 2.625 | 2820 | 1120 | 2780 | 760 | 2600 | 820 | 2380 | 780 | 2210 | 520 | 1860 | 570 |
| | 1 | 3.00 | 3.000 | 3240 | 1230 | 3200 | 830 | 2990 | 900 | 2710 | 850 | 2530 | 570 | 2120 | 620 |
| 3-1/2 | 1/2 | 7.00 | 1.750 | 1230 | 920 | 1210 | 640 | 1130 | 690 | 1160 | 650 | 1080 | 440 | 910 | 480 |
| | 5/8 | 5.60 | 2.188 | 1910 | 1070 | 1890 | 720 | 1760 | 780 | 1790 | 740 | 1660 | 500 | 1400 | 550 |
| | 3/4 | 4.67 | 2.625 | 2610 | 1200 | 2570 | 810 | 2400 | 870 | 2310 | 820 | 2150 | 560 | 1800 | 610 |
| | 7/8 | 4.00 | 3.063 | 3220 | 1320 | 3180 | 890 | 2970 | 960 | 2760 | 900 | 2570 | 610 | 2160 | 670 |
| | 1 | 3.50 | 3.500 | 3760 | 1440 | 3710 | 970 | 3460 | 1050 | 3170 | 990 | 2950 | 670 | 2480 | 730 |
| 4 | 1/2 | 8.00 | 2.000 | 1230 | 960 | 1210 | 700 | 1130 | 760 | 1160 | 720 | 1080 | 500 | 910 | 550 |
| | 5/8 | 6.40 | 2.500 | 1920 | 1220 | 1900 | 830 | 1770 | 900 | 1820 | 840 | 1690 | 570 | 1420 | 620 |
| | 3/4 | 5.33 | 3.000 | 2730 | 1370 | 2690 | 920 | 2510 | 1000 | 2520 | 940 | 2350 | 630 | 1970 | 690 |
| | 7/8 | 4.57 | 3.500 | 3520 | 1500 | 3470 | 1010 | 3240 | 1100 | 3090 | 1030 | 2880 | 700 | 2420 | 760 |
| | 1 | 4.00 | 4.000 | 4210 | 1640 | 4150 | 1110 | 3880 | 1200 | 3600 | 1130 | 3360 | 760 | 2820 | 830 |
| 4-1/2 | 5/8 | 7.20 | 2.813 | 1920 | 1320 | 1900 | 930 | 1770 | 1010 | 1820 | 950 | 1690 | 640 | 1420 | 700 |
| | 3/4 | 6.00 | 3.375 | 2760 | 1540 | 2730 | 1040 | 2550 | 1120 | 2610 | 1060 | 2440 | 710 | 2050 | 780 |
| | 7/8 | 5.14 | 3.938 | 3680 | 1690 | 3630 | 1140 | 3390 | 1240 | 3360 | 1160 | 3130 | 790 | 2630 | 860 |
| | 1 | 4.50 | 4.500 | 4560 | 1850 | 4500 | 1250 | 4200 | 1350 | 3990 | 1270 | 3710 | 860 | 3120 | 940 |
| | 1-1/4 | 3.60 | 5.625 | 6010 | 2160 | 5930 | 1460 | 5540 | 1580 | 5080 | 1490 | 4740 | 1000 | 3980 | 1100 |
| 5-1/2 | 5/8 | 8.80 | 3.438 | 1920 | 1330 | 1900 | 1010 | 1770 | 1090 | 1820 | 1030 | 1690 | 750 | 1420 | 820 |
| | 3/4 | 7.33 | 4.125 | 2770 | 1780 | 2730 | 1260 | 2550 | 1360 | 2620 | 1280 | 2440 | 870 | 2050 | 950 |
| | 7/8 | 6.29 | 4.813 | 3770 | 2060 | 3720 | 1400 | 3470 | 1510 | 3560 | 1420 | 3320 | 960 | 2790 | 1050 |
| | 1 | 5.50 | 5.500 | 4890 | 2260 | 4820 | 1520 | 4500 | 1650 | 4550 | 1550 | 4240 | 1050 | 3560 | 1150 |
| | 1-1/4 | 4.40 | 6.875 | 7020 | 2640 | 6930 | 1780 | 6470 | 1930 | 6110 | 1820 | 5690 | 1230 | 4780 | 1340 |
| 7-1/2 | 5/8 | 12.00 | 4.688 | 1920 | 1200 | 1890 | 950 | 1770 | 1030 | 1820 | 960 | 1690 | 730 | 1420 | 800 |
| | 3/4 | 10.00 | 5.625 | 2770 | 1720 | 2730 | 1320 | 2550 | 1430 | 2620 | 1340 | 2440 | 1010 | 2050 | 1110 |
| | 7/8 | 8.57 | 6.563 | 3770 | 2290 | 3720 | 1730 | 3470 | 1870 | 3560 | 1760 | 3320 | 1280 | 2780 | 1400 |
| | 1 | 7.50 | 7.500 | 4910 | 2870 | 4850 | 2060 | 4520 | 2230 | 4650 | 2100 | 4330 | 1430 | 3640 | 1560 |
| | 1-1/4 | 6.00 | 9.375 | 7680 | 3600 | 7580 | 2430 | 7070 | 2630 | 7260 | 2480 | 6770 | 1670 | 5680 | 1830 |
| 9-1/2 | 3/4 | 12.67 | 7.125 | 2770 | 1560 | 2730 | 1250 | 2550 | 1350 | 2620 | 1270 | 2440 | 970 | 2050 | 1060 |
| | 7/8 | 10.86 | 8.313 | 3770 | 2150 | 3720 | 1660 | 3470 | 1790 | 3560 | 1690 | 3320 | 1280 | 2790 | 1400 |
| | 1 | 9.50 | 9.500 | 4920 | 2800 | 4850 | 2130 | 4530 | 2300 | 4650 | 2170 | 4330 | 1630 | 3640 | 1780 |
| | 1-1/4 | 7.60 | 11.875 | 7680 | 4210 | 7580 | 3030 | 7070 | 3280 | 7270 | 3090 | 6770 | 2120 | 5690 | 2320 |
| | 1-1/2 | 6.33 | 14.250 | 11080 | 5250 | 10930 | 3540 | 10200 | 3830 | 10470 | 3610 | 9760 | 2440 | 8190 | 2660 |
| 11-1/2 | 7/8 | 13.14 | 10.062 | 3770 | 1950 | 3700 | 1590 | 3160 | 1730 | 3570 | 1630 | 3330 | 1230 | 2790 | 1350 |
| | 1 | 11.50 | 11.500 | 4920 | 2620 | 4860 | 2040 | 4530 | 2210 | 4650 | 2080 | 4330 | 1580 | 3640 | 1730 |
| | 1-1/4 | 9.20 | 14.375 | 7690 | 4130 | 7590 | 3140 | 7080 | 3400 | 7270 | 3200 | 6780 | 2380 | 5690 | 2600 |
| | 1-1/2 | 7.67 | 17.250 | 11070 | 5820 | 10920 | 4210 | 10190 | 4550 | 10470 | 4280 | 9750 | 2950 | 8190 | 3230 |
| 13-1/2 | 1 | 13.50 | 13.500 | 4920 | 2400 | 4850 | 1970 | 4540 | 2140 | 4670 | 2020 | 4350 | 1530 | 3650 | 1680 |
| | 1-1/4 | 10.80 | 16.875 | 7690 | 3940 | 7590 | 3030 | 7080 | 3280 | 7260 | 3080 | 6770 | 2340 | 5680 | 2560 |
| | 1-1/2 | 9.00 | 20.250 | 11070 | 5720 | 10930 | 4340 | 10200 | 4700 | 10460 | 4420 | 9750 | 3280 | 8190 | 3580 |

*Three (3) member joint.

By permission of NFPA

D. Connection with more than four members

    All equal thickness
        Evaluate single shear capacity of two of the members, and multiply by the total number of shear planes.

    Unequal thicknesses (this situation is complex and arises rarely)
- resolve into all possible adjacent 3-member joints
- evaluate capacity of each with above methods
- divide each capacity in half and assign to shear planes
- lesser of any two capacities applies for each plane
- connection capacity is multiple of number of shear and least capacity of those evaluated

### Load Distribution as Influenced by Connection Details

Connection details can alter the way in which loads are introduced into members through connections. In the following example, a simple truss is used to demonstrate the crucial effect that connection details have on the angle of load application with the grain.

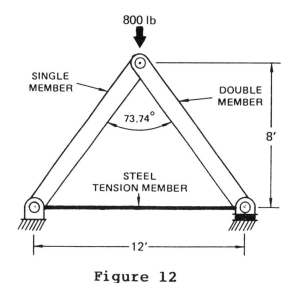

**Figure 12**
**Entire Truss, as Loaded**

The 800 pound load can be imposed on the fastener, the single member, or the doubled members, depending upon how the connection at the top is detailed. In all cases, the timber members are simple compression members; the difference is in the angles of load application at the fasteners. In this and all multi-member connections, dividing the connection up into pieces and investigating each piece as a free body is the best way to analyze the load distribution.

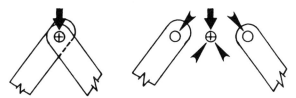

**Figure 13**
**Load Applied Through Bolt**

If the load is applied to the bolt as shown in figure 13, both pieces act as simple truss members, and the fastener load is parallel to the grain in all members.

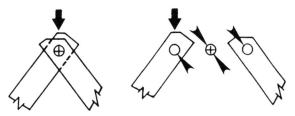

**Figure 14**
**Load Applied Through Single Member**

In figure 14, the doubled members are two-force bodies or simple truss members. Therefore, the fastener load is parallel to the grain in the doubled members. In the single member, however, the load applied by the doubled members through the fasteners is at an angle of 73.74° with the grain.

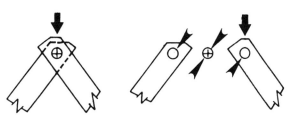

**Figure 15**
**Load Applied Through Doubled Members**

The load application shown in figure 15 gives an opposite response to those of figure 14. The doubled members have the fastener load parallel to the grain. The fastener load in the single member is at an angle of 73.74° with the grain.

### G. Spacing Requirements for Bolts and Lag Screws

The design values for both lag screws and bolts are valid only if certain minimum spacing requirements are met. Some of the spacings can be reduced to absolute minimums, with a reduced design capacity calculated by linear interpolation to zero capacity at zero spacing. A row is defined as two or more fasteners in a line parallel to the load direction. All spacings are given as center to center distances, and are multiples of the connector diameter.

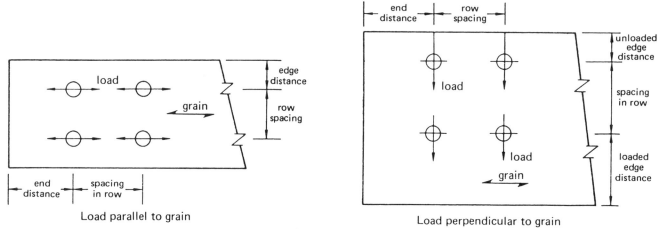

**Figure 16**
**Spacing Dimensions -- Bolts and Lag Screws**

The following is an outlined form of the required spacings for bolts and lag screws. The spacings will vary with load direction. Some spacings may be reduced to an absolute minimum value with a linear decrease in capacity. See figure 10 for an explanation of staggered rows of bolts.

<u>Minimum Full-Strength Spacings</u>

    Spacing of fasteners in a row
        Load parallel to grain
            4 diameters for full-strength,
            3 diameters absolute minimum (75% bolt design value)
        Load perpendicular to grain
            Spacing limited by spacing requirements of attached
            members (wood or metal), loaded parallel to grain

Spacing of fastener rows

    Load parallel to grain

        1-1/2 diameters, across the grain

    Load perpendicular to grain; function of fastener L/d ratio

        $L/d = 2$; 2-1/2 diameters, along grain

        $L/d \geq 6$; 5 diameters, along grain

        $2 \leq L/d \leq 6$; linear interpolation between 2-1/2 and 5 diameters

    NOTE: In any case, if the row spacing parallel to the member exceeds 5 inches, separate splice plates are required for each row.

End distance

    Load parallel to grain

        Tension

            Softwood: 7 diameters, 3-1/2 absolute minimum (50% design value)

            Hardwood: 5 diameters, 2-1/2 absolute minimum (50% design value)

        Compression

            4 diameters, 2 diameters absolute minimun (50% design value)

    Load perpendicular to grain

        4 diameters, 2 absolute minimum (50% design value)

Edge distance (L = live bolt length, exclusive of enclosed member thickness)

    Load parallel to grain; function of fastener L/d ratio

        $L/d \leq 6$; 1-1/2 diameters

        $L/d > 6$

            **greater** of:
- 1-1/2 diameters
- 1/2 distance between fastener rows

    Load perpendicular to grain

        Loaded edge; 4 diameters

        Unloaded edge; 1-1/2 diameters

    NOTE: Loads should not be hung from beams on fasteners below the neutral axis.

When loads are at angles other than parallel and perpendicular to the grain, the engineer should ensure uniform load distribution by aligning the center of the connection resistance with the centroids of the connected members. The spacing constraints should be met as well as is feasible in the members, recognizing that these situations can be complex and even conflicting.

## Example 7

The drawing below shows five feasible bolt patterns using six 5/8" bolts to splice a 4x6 member with two 2x6 splice plates. The members will be in tension. Determine the capacity and the minimum splice plate lengths for each pattern. All members are Southern Pine #1, surfaced and used at 15% moisture content. Design will be for normal load duration.

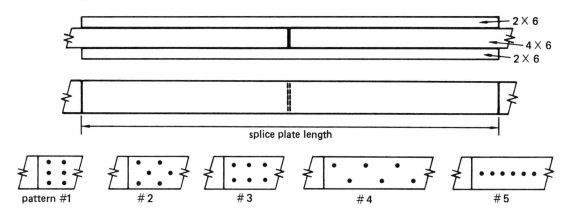

[NOTE: In designing tension splices, the engineer is trying to maximize the load capacity while keeping the splice length within reasonable limits. This optimization process often involves balancing connector and member capacities. There are several variables to consider: the distance between bolts in a row, the row spacing, the row staggering, and the end and edge distances. These dimensions influence the numbers of connector rows and connectors within those rows that the connection is considered as having in the connector capacity analysis. The layout also governs the number of connectors considered to be within a given net cross section in the member capacity analysis. The purpose of this example is to demonstrate the interaction of these design variables in optimizing the design.

Check the capacity of the bolts, correcting for multiple connectors. Make sure the net section left after drilling does not limit load capacity. Use required spacing and end distances to determine the minimum length splice members.

The allowable tension stress in the members and basic allowable load in the bolts do not change with the bolt pattern. To find the basic bolt capacity, use table 13 with the correct member thickness. This is double shear with aligned members, but the member thickness must be compared to determine which controls.

$$t_{\text{side member}} = 1.5" < (.5)(t_{\text{main}}) = (.5)(3.5") = 1.75"$$

Therefore, use $t = (2)(t_{\text{thinnest side member}}) = 3"$.

With Southern Pine, 5/8" bolt, and load parallel to grain, table 13 gives P = 1960 lb/bolt.

The stress tables of appendix 2 give an allowable tension stress of 1050 psi. Note that with the #1 grade and 6" width, footnote 3 of that table does not decrease the allowable stress.

Table 4 gives the CUF = 1.0 for this protected exposure condition. The LDF is 1.0 (table 2).

The area of the main member is 19.25 in$^2$, and that of each side plate is 8.25 in$^2$, before drilling any holes. The holes are assumed to be 1/16" oversized, or 11/16" diameter. The minimum spacings are given as multiples of the bolt diameter. For the 5/8" bolts, the minimum spacings are:

| | | |
|---|---|---|
| End Distance (tension) | 7 diameters | 4.375" |
| Edge Distance | 1 1/2 dia. | 15/16" |
| Row Spacing | 1 1/2 dia. | 15/16" |
| Spacing in Row | 4 diameters | 2.5" |

The edge spacing varies with the L/d ratio. The 4x6 is enclosed, and its thickness does not contribute to live bolt length (L= 2 x 1-1/2 = 3"). In this case, L/d = 3"/(5/8") = 4.8 < 6, so that the 1-1/2 diameters controls.

With these common numbers, start evaluating the individual capacities.

BOLT PATTERN #1:

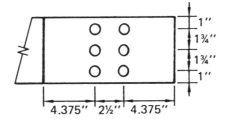

There is adequate space across the member for the three rows of two bolts. The minimum splice plate length is

(2) [(2.5") + (2)(4.375")] = 22.5"

Check the bolt capacity first. This pattern is considered as 3 rows of 2 bolts each. Table 8A shows a K of 1.0 for 2 bolts, no matter what the member sizes are. Therefore, the bolt capacity is

(6 bolts) (1960 lb/bolt) (1.0) = 11,760 lb

Now check the net section capacity. A 1/8 reduction in area is made to account for the possibility of knots also falling in the critical cross section. The splice plates will control because they are thinner than the main member. The area of 3 holes will be removed.

Net Area = (2) [(8.25 in$^2$) - (3)(1.5")(11/16")] (7/8) = 9.02 in$^2$

The net section capacity, therefore, is

(9.02 in$^2$) (1050 psi) = 9,475 lb

For bolt pattern #1, the member controls the capacity at 9,475 pounds.

Since the bolts will not be loaded to capacity when the splice plates are, the bolt spacing can be reduced somewhat. The bolts will be at 9475/11760 = .81, or 81% of their design capacity. The bolt spacing could be reduced to 2" (3 diameters is the absolute minimum allowed), resulting in an absolute minimum splice plate length of 21.5".

BOLT PATTERN #2:

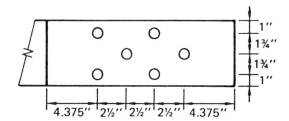

Since the member controls the capacity in pattern #1, we will stagger the pattern to increase the net section. The in-row spacing must be at least 8 bolt diameters while keeping S < 4d. If S $\geq$ 4d this pattern would be considered as 1 row of 4 bolts and 1 row of 2 bolts.

The minimum splice plate length is
  (2) [(3)(2.5") + (2)(4.375")] = 32.5"

Check the bolt capacity first. This pattern is still considered as 3 rows of 2 bolts each (S < 4d). Therefore, the bolt capacity is
  (6 bolts) (1960 lb/bolt) (1.0) = 11,760 lb

Now check the net section capacity. With this staggered spacing there is a maximum of only 2 bolts in any 4 diameter (2.5") length of the members. Therefore, only 2 holes will be subtracted from the area.

Net Area = (2) [(8.25 in$^2$) - (2)(1.5")(11/16")] (7/8) = 10.83 in$^2$

The net section capacity, therefore, is

(10.83 in$^2$) (1050 psi) = 11,370 lb

For bolt pattern #2, the member controls the capacity at 11,370 pounds. The bolt and member capacities are very nearly balanced, making this a reasonable design choice.

BOLT PATTERN #3:

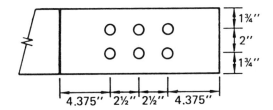

The minimum splice plate length is

(2) [(2)(2.5") + (2)(4.375")] = 27.5"

Check the bolt capacity first. This pattern is considered as 2 rows of 3 bolts each. The multiple fastener factor, K, with wood side plates, is found in table 8A. To use that table, evaluate the ratio of the member areas.

$A_1/A_2$ = (19.25 in$^2$)/(2)(8.25 in$^2$) = 1.17

Since this is greater than 1.0, use the reciprocal, 1/1.17 = .86

When using this reciprocal, also use $A_2$ = (2)8.25 in$^2$) = 16.5 in$^2$, which falls in the 12-19 in$^2$ range.

Interpolating between 0.95 for a 0.5 ratio, and 0.97 for a 1.0 ratio, the K is 0.97 in this case.

Therefore, the bolt capacity is

(6 bolts)(1960 lb/bolt)(0.97) = 11,410 lb

Now check the net section capacity. Only 2 holes will be subtracted from the area.

Net Area = (2)[(8.25 in²) − (2)(1.5")(11/16")](7/8) = 10.83 in²

The net section capacity, therefore, is

(10.83 in²)(1050 psi) = 11,370 lb

For bolt pattern #3, the member controls the capacity at 11,370 pounds. This is another well balanced connection design, with an even shorter length than the previous layout.

BOLT PATTERN #4:

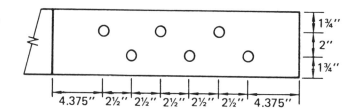

The minimum splice plate length is

(2)[(5)(2.5") + (2)(4.375")] = 42.5"

Check the bolt capacity first. This pattern is considered as 2 rows of 3 bolts each. Again because S < 4d, the multiple connector factor is the same for bolt pattern #3. Therefore, the bolt capacity is

(6 bolts)(1960 lb/bolt)(0.97) = 11,410 lb

Check the net section capacity. Because only one hole occurs in any 4 diameter (2.5") length of the members, the area of only one hole need be subtracted.

Net Area = (2)[(8.25 in²) − (1)(1.5")(11/16")](7/8) = 12.63 in²

The net section capacity, therefore, is

(12.63 in$^2$) (1050 psi) = 13,260 lb

For bolt pattern #4, the bolts control the capacity at 11,410 pounds. This capacity is slightly larger than that of #3, but probably not enough to justify the extra 15 inches of splice plate length.

BOLT PATTERN #5:

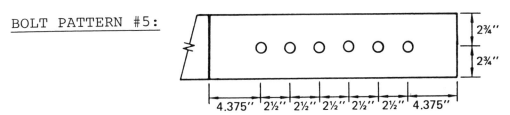

The minimum splice plate length is still

(2) [(5)(2.5") + (2)(4.375")] = 42.5"

Check the bolt capacity first. This pattern is considered as 1 row of 6 bolts. The multiple connector factor is found with the same relative areas used in bolt pattern #3, except for 6 bolts. Now the interpolation for $A_2/A_1$ = 0.86 is between K = 0.75 and 0.84. That interpolation gives a K = 0.81. Therefore, the bolt capacity is

(6 bolts) (1960 lb/bolt) (0.81) = 9,530 lb

Check the net section capacity. The area of only one hole need be subtracted.

Net Area = (2) [(8.25 in$^2$) − (1)(1.5")(11/16")] (7/8) = 12.63 in$^2$

The net section capacity, therefore, is

(12.63 in$^2$) (1050 psi) = 13,260 lb

For bolt pattern #5, the bolts control the capacity at 9,530 pounds. This is almost as low as the capacity of bolt pattern #1. Note that bolt patterns #4 and #5 differ only in the amount of staggering. The minimum amount of row stagger which would make the 6 bolts act as two separate rows is the along the grain spacing divided by four, or 5/8".

Considering splice plate length and balanced capacity, either bolt pattern #3 or #4 would be a good design.

## Example 8

What is the capacity of this connection against snow load? The spruce-pine-fir seasons in place, and it is expected to stay dry.

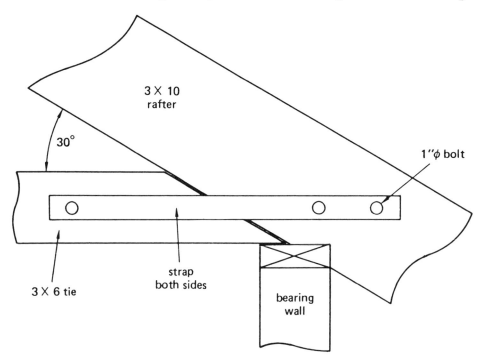

Check the tie bolt capacity and the capacity of the bolts in the rafter. Use the Hankinson formula to determine the rafter bolts' capacity.

Table 2 gives a LDF of 1.15 with snow loads. According to table 4, the CUF for this connection is 1.0. There will be a 75% increase in the bolt capacity, parallel to the grain only, because of the metal side plates. Table 8B shows that the multiple fastener factor, K, for any connection with only two fasteners in a row is 1.0.

First, evaluate the capacity in the tie beam. Table 13, with spruce-pine-fir, 1" bolts, and t = 2.5", gives P = 2260 lb/bolt. The bolt capacity is

(1) bolt (2260) lb/bolt (1.15) (1.0) (1.75) (1.0) = 4550 lb.

When evaluating the bolt capacity in the rafter, the CUF of table 9 must be included because the connection seasons in place. The strap does cross the grain, so the CUF of table 9 is 0.4. Table 13 gives a capacity perpendicular to the grain, Q, of 700 lb. Before using the Hankinson formula, find the adjusted bolt capacities in both directions.

$$P = (2260) \text{ lb } (1.15)(1.75)(0.4) = 1819 \text{ lb/bolt}$$
$$Q = (700) \text{ lb } (1.15)(0.4) = 322 \text{ lb/bolt}$$

Note that only the component parallel to the grain is increased 75% for the metal side plates (See p. 47, item C.2). Using the Hankinson formula,

$$F_{30°} = \frac{(322) \text{ lb } (1819) \text{ lb}}{(1819) \text{ lb } \sin^2 30° + (322) \text{ lb } \cos^2 30°} = 841 \text{ lb/bolt}$$

The rafter connection capacity is (2) bolts (841) lb/bolt = 1680 lb

The two bolts in the rafter limit the connection capacity to 1680 lb, significantly less than the capacity of the single bolt in the tie beam.

The connection was weakened most by having the timbers season in place with the steel strap bolted across the grain. Since the strap is not exactly across the grain, but at an angle to it, some less severe CUF might be justified. One possibility would be to use the Hankinson formula to interpolate between the 1.0 and 0.4 factors (CUF = 0.73). The rafter connection capacity would then be:
    (2) bolts (1536) lb/bolt = 3072 lb

### H. Shear Plates and Split Rings (Connectors)

Shear plates and split rings are highly efficient mechanical fasteners which can have capacities five to six times greater than those of bolts or lag screws. Shear loads are transmitted between members through the metal connectors which fit snugly into carefully cut grooves or daps. The bolt or lag screw which lies on the connector axis serves only to hold the joint together. Both connector types are manufactured by the Timber Engineering Company (TECO) whose design charts and tables form the basis of this section.

Split rings come in 2 1/2" and 4" diameters, and provide wood-to-wood connection. Lag screws are used to hold a split ring connection together when the holding member is very large or has an inaccessible far side. Shear plates have 2 5/8" or 4" diameters. Shear plates are used singly to attach a metal side member, or in pairs back-to-back, to connect two wood members. In either case, lag screws can be used with shear plates for the same reasons they are used with split rings. Figure 17 illustrates the sizes and types of timber connectors.

Split ring

Pressed steel shear plate

Malleable iron shear plate

**Figure 17**
Types of Split Ring and Shear Plates
By permission of TECO

Considerations in Connector Design

Most of the factors applied to simpler fasteners also apply to connectors. The load duration factors are the same. The condition of use factors specific to connectors are found in table 4. The modifier for number of connectors in a row is found in tables 8A and 8B. The TECO capacity tables are specific to angle of load and grain, and already include the Hankinson formula interpolation. There is an absolute maximum load for the shear plates established by the bolt bearing on the plate. This maximum capacity is found in the lower right hand corner of the shear plate capacity tables.

There are several special design considerations specific to shear plates and split rings.

- Lag Screws in Connectors: If lag screws are used instead of bolts, the factors in table 14 apply.

- Net Sections: Net sections are determined by subtracting the projected area of the cut daps and bolt holes. This calculation is aided by the tabulated projected connector areas of appendix 4.

A bolt hole should have a diameter no larger than 1/16" plus the bolt diameter. Half-inch bolts and lag screws are used in 2 1/2" split rings, while 3/4" bolts and lag screws are used in all the other connectors.

Because a knot near a connector would further reduce the net section, the allowable stresses in tension and compression are reduced by 1/8. This reduction is applied only to solid sawn lumber, not to glue-laminated members.

Connectors in adjacent staggered rows are treated as being in the same cross section if the parallel to the grain spacing between them is less than four connector diameters.

### Table 14
### Capacity Modification Using Lag Screws in Connectors

Factors apply to design values tabulated for connector units used with bolts

| Connector size and type | Side plate | Pene-tration | Penetration of lag screw into member receiving point (Number of shank diameters) Fastener species group (See Table 3) | | | | Modification factor[1] |
|---|---|---|---|---|---|---|---|
| | | | I | II | III | IV | |
| 2-1/2" split ring 4" split ring 4" shear plate | Wood or Metal | Standard | 7 | 8 | 10 | 11 | 1.00 |
| | | Minimum | 3 | 3-1/2 | 4 | 4-1/2 | 0.75 |
| 2-5/8" shear plate | Wood | Standard | 4 | 5 | 7 | 8 | 1.00 |
| | | Minimum | 3 | 3-1/2 | 4 | 4-1/2 | 0.75 |
| 2-5/8" shear plate | Metal | Standard and Minimum | 3 | 3-1/2 | 4 | 4-1/2 | 1.00 |

1. Use straight line interpolation for intermediate penetrations.

By permission of NFPA

## Example 9

What is the net section of the timber prepared as shown for a connector?

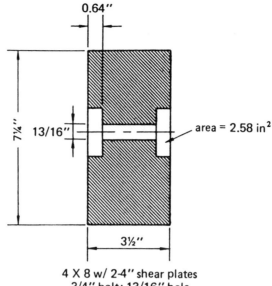

4 X 8 w/ 2-4" shear plates
3/4" bolt; 13/16" hole

Appendix 5 gives the 25.375 in² gross area for the 4x8. Appendix 4 lists the projected area of a 4" shear plate as 2.58 in². After subtracting two shear plates, the rest of the bolt hole between the plates must be subtracted. Appendix 4 also gives the depth of the connectors, .64" for a 4" shear plate. The 3/4" bolt has a 13/16" bolt hole. The length of the hole between the plates is 3.5" - (2)(.64") = 2.22 in.

The net area then is:

$$25.375 \text{ in}^2 - (2)(2.53) \text{ in}^2 - (13/16")(2.22") = 18.51 \text{ in}^2$$

- **Minimum Lumber Sizes:** TECO recommends minimum lumber widths and thicknesses in order to avoid members being unduly weakened by the cut daps and grooves. The minimum sizes depend on the connector type, diameter, and whether a member has connectors in one or two faces on any one bolt. These minimum sizes are found in the appropriate connector tables of figures 21 through 30.

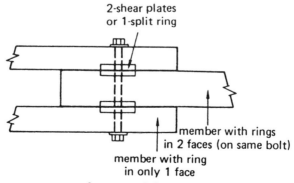

**Figure 18**
**Number of Faces with Connectors**

- Connector Spacing: The connector design capacities are based on certain minimum spacings being maintained. Some of these spacings may be reduced, with a proportionate reduction in the connector capacity. The spacing charts in figures 21 through 30 allow for simultaneous adjustment of spacings both along and across the grain. The distances are still measured from center to center of connectors. These charts incorporate load angles and reductions for underloading. They are powerful detailing aids once their use is understood.

Edge Distance: The edge distance charts give the minimum required loaded edge distances on the horizontal axis. The percentage of full load is the vertical axis. For loads parallel to the grain, and for any unloaded edge, the minimum edge distance is the value on the right-hand border of the chart. For any other load-grain angle, the diagonal lines represent the acceptable minimum edge distances as a function of percentage of full load.

End Distance: The end distance charts are very similar to the edge distance charts. The tension line is used for a load with any component toward the near end of the member. For members in compression, there is a family of lines for various load-grain angles.

End Distances Without Square-cut Ends: There are two limitations on end distance in members which are not cut off square. The two can be converted to equivalent dimensions and compared with each other to determine which one controls.

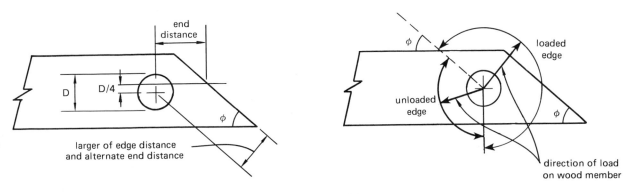

**Figure 19**
**End Distance in Members Without Square Ends**

To determine which of the limitations controls, find the larger of these two values:

- Loaded or unloaded edge distance (see figure 19 for choice)

- (tabulated end distance)$(\sin\phi) + (D/4)(\cos\phi)$         (3)

The longer length, when laid out normal to the angled cut, is the outer limit of acceptable connector positions.

Connector Spacing: These charts deal with the spacing of any two connectors in the same connection. The grid presents all possible options of spacing across and along the grain. This gives the engineer much flexibility in making the overall connection fit within the limited space of two intersecting members, while still meeting all spacing restrictions.

The vertical axis is the across the grain spacing of any two connectors, while the horizontal axis has a range of along the grain spacings. The lines radiating from the origin represent the angle that the line connecting the centers of two connectors makes with the grain, if they are spaced according to the grid position. This should not be confused with the angle that the load makes with the grain.

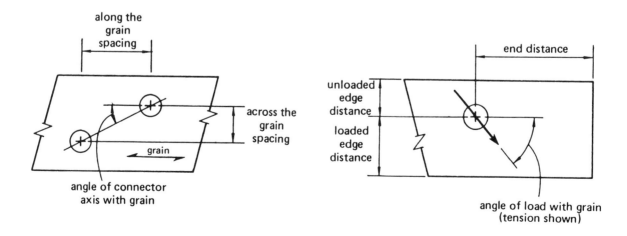

**Figure 20**
**Connector Spacing and Angle to the Grain**

# Table 15
## Species Groupings for Shear Plates and Split Rings

| Connector load group* | Grouping for timber connector loads — Species of wood |
|---|---|
| Group A | Ash, Commercial White<br>Beech<br>Birch, Sweet & Yellow<br>Douglas Fir-Larch (Dense)***<br>Hickory & Pecan<br>Maple, Black & Sugar<br>Oak, Red & White<br>Southern Pine (Dense) |
| Group B | Douglas Fir-Larch***<br>Southern Pine<br>Sweetgum & Tupelo |
| Group C | California Redwood<br>Douglas Fir, South<br>Eastern Hemlock-Tamarack***<br>Eastern Spruce<br>Hem - Fir***<br><br>Lodgepole Pine<br>Mountain Hemlock<br>Northern Aspen<br>Northern Pine<br>Ponderosa Pine****<br>Ponderosa Pine-Sugar Pine<br>Red Pine****<br>Sitka Spruce<br>Southern Cypress<br><br>Spruce-Pine-Fir<br>Western Hemlock<br>Yellow Poplar |
| Group D | Aspen<br>Balsam Fir<br>Black Cottonwood<br>California Redwood, Open grain<br>Coast Sitka Spruce<br><br>Cottonwood, Eastern<br>Eastern White Pine***<br><br>Engelmann Spruce - Alpine Fir<br>Idaho White Pine<br><br>Northern White Cedar<br><br>Western Cedars***<br>Western White Pine |

*When stress graded.
**Based on weight and volume when oven-dry.
***Also applies when species name includes the designation "North".
****Applies when graded to NLGA rules.

Note: Coarse grain Southern Pine, as used in some glued laminated timber combinations, is in Group C.

By permission of NFPA

The parabolic curves on the connector spacing charts are specific to the angle that the load makes with the grain. Each parabola passes through all the acceptable combinations of spacings across and along the grain for that load-grain angle. The quarter circle represents all the acceptable spacing combinations, no matter what angle the load makes with the grain, if the connector is loaded to only 50% of its design capacity.

There are two interpolations used with this chart; for angles of load between those given, and for load levels between full and half capacity.

- <u>Species Grouping</u>: For convenience, the species are still grouped into capacity ranges. The grouping is slightly different from the one used with the other fasteners, and is found in table 15.

- <u>Connectors in End Grain</u>: Connectors are installed in the end grain of timber very rarely. For a rather complicated treatment of this difficult subject, refer to the National Design Specifications.

## Example 10

What is the snow load capacity of the splice? Check connector and net section capacities, and spacings. The splice is on the lower chord of a truss that was fabricated wet and is always dry in service.

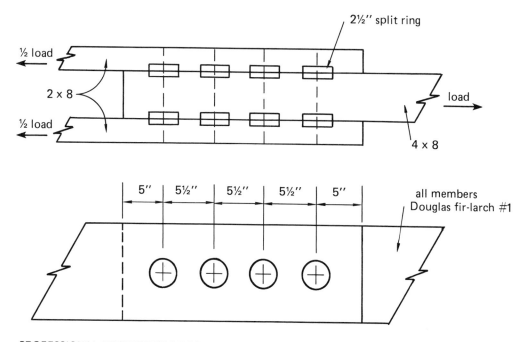

# Timber Design

Check the connector capacity using the multiple connector correction. Then, check the net section capacity after subtracting the connector projected area. Use appropriate design values for the timber and any necessary footnoted considerations. Finally, check that the spacing is at least the minimum required for the members, connectors, and load levels involved.

The snow load has a LDF of 1.15 (table 2). Table 4 gives a CUF of 0.8. According to table 15, Douglas fir-larch is a group B species.

The split ring capacities must be looked up for the main members and the splice plates. Figure 21 has both capacities. In the 4x8, using the 2" THICK-2 FACES curve, the capacity at $0°$ to the grain is 2740 lb. Even though the member is actually 3.5" thick, the curve for 2" thick members is used as the highest available. In the 2x8 splice plates, using the 1 1/2"THICK-1 FACE curve, the capacity along the grain is 2740 lb.

The multiple connector factor, K, of figure 8A is evaluated next. The ratio $A_1/A_2$, with information from appendix 5, is

$$(25.375) \text{ in}^2 / (2)(10.875) \text{ in}^2 = 1.167.$$

This ratio is greater than 1.0. Therefore, use $A_2/A_1 = 1/1.167 = 0.857$, and use $A_2$ instead of $A_1$ to get K. The K for $A_2/A_1 = .5$ is .93; and for $A_2/A_1 = 1.0$, K is .97. Interpolating for the actual $A_2/A_1$, K = .96.

Connector capacity = (8) connectors (2740) lb/connector (.96)(1.15)(.8)
= 19,360 lb

The net section capacities are determined in both the main member and the splice plates. The 4x8 net area is

$$25.375 \text{ in}^2 - (2)(1.10) \text{ in}^2 - (9/16")[3.5"-(2)(.375")] = 21.63 \text{ in}^2.$$

Appendix 4 yields the 1.10 in$^2$ area for the connectors and the .375" depth of the connector. The 1/2" bolt has a 9/16" hole that must be subtracted between the connectors.

The 2x8's net area is

$(2)(10.875)$ in$^2$ − $(2)(1.10)$ in$^2$ − $(9/16")[(2)(1.5") − (2)(.375")]$ = 18.28 in$^2$.

This net area is less than the main members. With equal allowable stresses, the stress in the splice plates will control.

The allowable tension is

$F_t$ = (1000) psi (.8) (7/8) = 700 psi.

The .8 factor comes from footnote 3, appendix 2. The 1/8 reduction is recommended for critical sections, in order to account for possible knots.

The member capacity, as limited by the two 2x8's, is

(18.28) in$^2$ (700) psi (1.15) = 14,715 lb.

The splice capacity is, therefore, limited by net section stresses in the splice plates to 14,715 lb.

The spacing requirements are checked against the requirements in figure 22. The minimum member width with 2 1/2" split rings is 3.5", less than the actual 7.25". The minimum member thickness, with 2 1/2" split rings, is 1" with connectors in one face and 1.5" with connectors in two faces. These are both less than provided, and therefore adequate.

When the splice plates are at capacity, the connectors are only loaded to (14,715)/(19,360) = .76, or 76%. The end distance chart in figure 22 shows that the minimum required is 4", again less than provided.

The minimum spacing (parallel to grain) is found in the spacing chart of figure 22. With the 76% load, interpolate between the 50% value of 3.5" and the 6.75" required for full strength.

$$3.5" + \frac{76\% - 50\%}{100\% - 50\%} (6.75" - 3.5") = 5.19" < 5.5" \text{ provided OK}$$

All the spacing provided is adequate for loads applied to connectors.

# TIMBER DESIGN

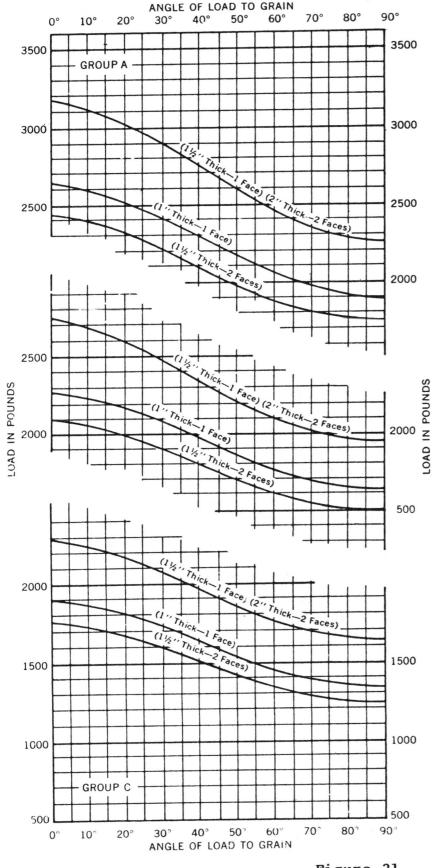

## 2½" SPLIT RING DATA

| Split Ring—Dimensions | |
|---|---|
| Inside Diameter at center when closed | 2½" |
| Inside diameter at center when installed | 2.54" |
| Thickness of ring at center | 0.163" |
| Thickness of ring at edge | 0.123" |
| Depth | ¾" |
| **Lumber, Minimum dimensions allowed** | |
| Width | 3½" |
| Thickness, rings in one face | 1" |
| Thickness, rings opposite in both faces | 1½" |
| **Bolt, diameter** | ½" |
| Bolt hole, diameter | 9/16" |
| **Projected Area for portion of one ring within a member, square inches** | 1.10 |
| **Washers, minimum** | |
| Round, Cast or Malleable Iron, diameter | 2⅝" |
| Square Plate Length of Side | 2" |
| Thickness | ⅛" |
| (For trussed rafters and similar light construction standard wrought washers may be used.) | |

### SPLIT RING SPECIFICATIONS

Split rings shall be TECO split rings as manufactured by TECO, Washington, D.C. Split rings shall be manufactured from hot rolled S. A. E.—1010 carbon steel. Each ring shall form a closed true circle with the principal axis of the cross section of the ring metal parallel to the geometric axis of the ring. The ring shall fit snugly in the prepared groove. The metal section of each ring shall be beveled from the central portion toward the edges to a thickness less than that at mid-section. It shall be cut through in one place in its circumference to form a tongue and slot.

### PERCENTAGES FOR DURATION OF MAXIMUM LOAD

| | |
|---|---|
| Two Months Loading, as for snow | 115% |
| Seven Days Loading | 125% |
| Wind or Earthquake Loading | 133⅓% |
| Impact Loading | 200% |
| Permanent Loading | 90% |

### DECREASES FOR MOISTURE CONTENT CONDITIONS

| Condition when Fabricated | Seasoned | Unseasoned | Unseasoned |
|---|---|---|---|
| Condition when Used | Seasoned | Seasoned | Unseasoned or Wet |
| Split Rings | 0% | 20% | 33% |

By permission of TECO

**Figure 21**
Design and Load Data for TECO Connectors   2 1/2" TECO Split Rings

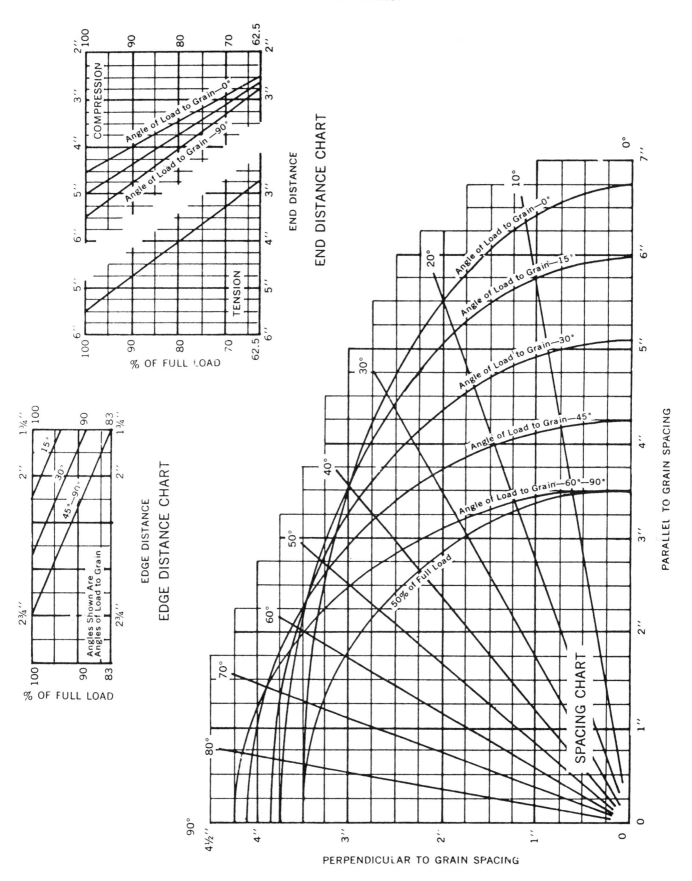

**Figure 22** By permission of TECO
Design and Load Data for TECO Connectors 2 1/2" TECO Split Rings

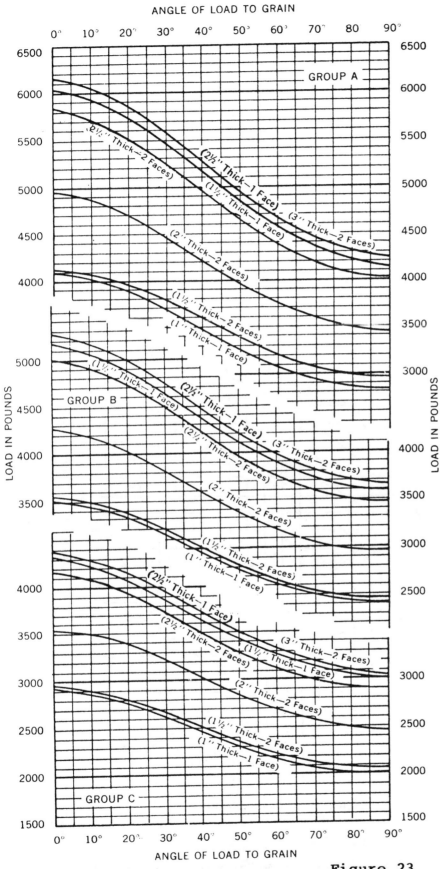

## 4" SPLIT RING DATA

| Split Ring—Dimensions | |
|---|---|
| Inside Diameter at center when closed | 4" |
| Inside diameter at center when installed | 4.06" |
| Thickness of ring at center | 0.193" |
| Thickness of ring at edge | 0.133" |
| Depth | 1" |
| Lumber, Minimum dimensions allowed | |
| Width | 5½" |
| Thickness, rings in one face | 1" |
| Thickness, rings opposite in both faces | 1½" |
| Bolt, diameter | ¾" |
| Bolt hole, diameter | 13/16" |
| Projected Area for portion of one ring within a member, square inches | 2.24 |
| Washers, minimum | |
| Round, Cast or Malleable Iron, diameter | 3" |
| Square Plate | |
| Length of Side | 3" |
| Thickness | 3/16" |
| (For trussed rafters and similar light construction standard wrought washers may be used.) | |

### SPLIT RING SPECIFICATIONS

Split rings shall be TECO split rings as manufactured by TECO, Washington, D.C. Split rings shall be manufactured from hot rolled S. A. E.—1010 carbon steel. Each ring shall form a closed true circle with the principal axis of the cross section of the ring metal parallel to the geometric axis of the ring. The ring shall fit snugly in the prepared groove. The metal section of each ring shall be beveled from the central portion toward the edges to a thickness less than that at mid-section. It shall be cut through in one place in its circumference to form a tongue and slot.

### PERCENTAGES FOR DURATION OF MAXIMUM LOAD

| | |
|---|---|
| Two Months Loading, as for snow | 115% |
| Seven Days Loading | 125% |
| Wind or Earthquake Loading | 133⅓% |
| Impact Loading | 200% |
| Permanent Loading | 90% |

### DECREASES FOR MOISTURE CONTENT CONDITIONS

| Condition when Fabricated | Seasoned | Unseasoned | Unseasoned |
|---|---|---|---|
| Condition when Used | Seasoned | Seasoned | Unseasoned or Wet |
| Split Rings | 0% | 20% | 33% |

By permission of TECO

**Figure 23**
Design and Load Data for TECO Connectors 4" TECO Split Rings

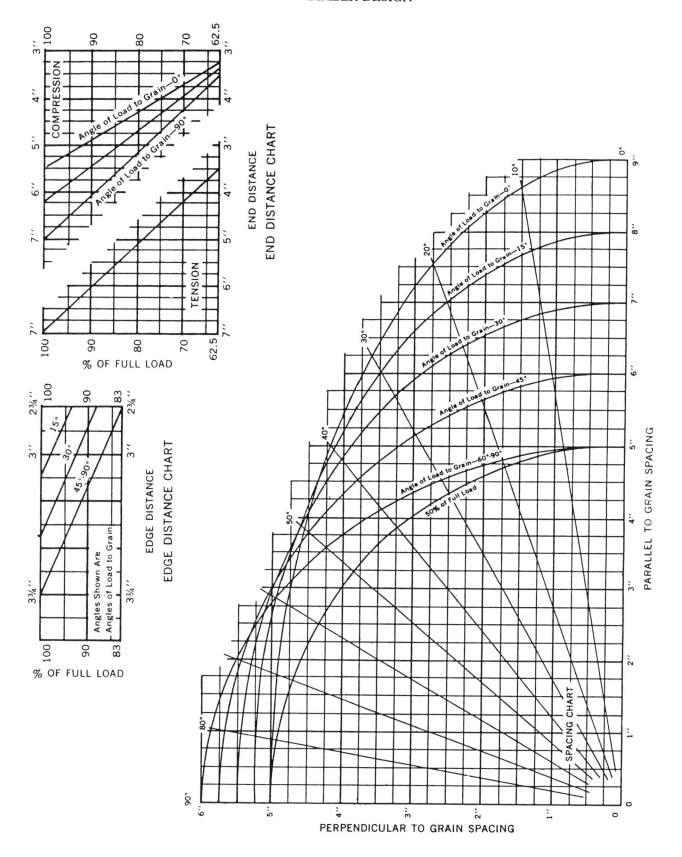

**Figure 24** By permission of TECO
Design and Load Data for TECO Connectors 4" TECO Split Rings

# Timber Design

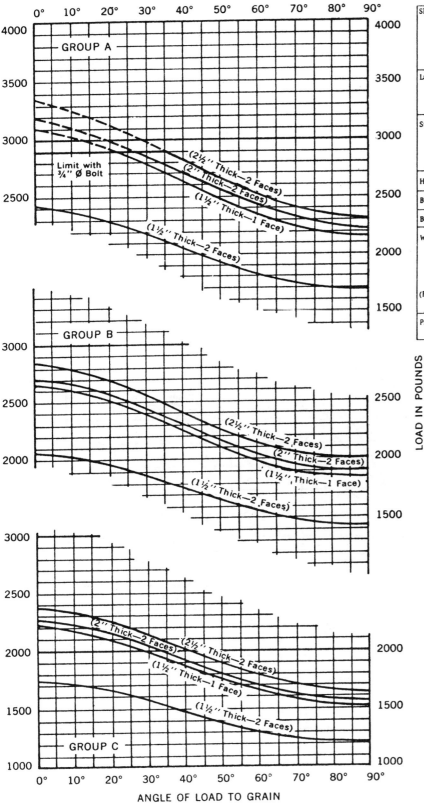

## 2 5/8" SHEAR PLATE DATA

| Shear Plates, Dimensions | Pressed Steel | |
|---|---|---|
| Material | Reg. | Lt. Ga. |
| Diameter of plate | 2.62" | 2.62" |
| Diameter of bolt hole | .81" | .81" |
| Depth of plate | .42" | .35" |
| Lumber, minimum dimensions | | |
| Face, width | 3½" | 3½" |
| Thickness, plates in one face only | 1½" | 1½" |
| Thickness, plates opposite in both faces | 1½" | 1½" |
| Steel Shapes or Straps (Thickness required when used with shear plates) Thickness of steel side plates shall be determined in accordance with A.I.S.C. recommendations. | | |
| Hole, diameter in steel straps or shapes | 13/16" | 13/16" |
| Bolt, diameter | 3/4" | 3/4" |
| Bolt Hole, diameter in timber | 13/16" | 13/16" |
| Washers, standard, timber to timber connections only | | |
| Round, cast or malleable iron, diameter | 3" | 3" |
| Square Plate | | |
| Length of side | 3" | 3" |
| Thickness | 1/4" | 1/4" |
| (For trussed rafters and other light structures standard wrought washers may be used.) | | |
| Projected Area, for one shear plate, square inches | 1.18 | 1.00 |

### SHEAR PLATE SPECIFICATIONS

Shear Plates shall be TECO shear plates as manufactured by TECO, Washington, D.C. Pressed Steel Type—Pressed steel shear-plates shall be manufactured from hot-rolled S.A.E.—1010 carbon steel. Each plate shall be a true circle with a flange around the edge extending at right angles to the face of the plate and extending from one face only, the plate portion having a central bolt hole and two small perforations on opposite sides of the hole and midway from the center and circumference.

### PERCENTAGES FOR DURATION OF MAXIMUM LOAD

| | |
|---|---|
| Two Months Loading, as for snow | *115% |
| Seven Days Loading | *125% |
| Wind or Earthquake Loading | *133⅓% |
| Impact Loading | *200% |
| Permanent Loading | 90% |

* Do not exceed limitations for maximum allowable loads for shear plates given elsewhere on this page.

### DECREASES FOR MOISTURE CONTENT CONDITIONS

| Condition when Fabricated | Seasoned | Unseasoned | Unseasoned |
|---|---|---|---|
| Condition when Used | Seasoned | Seasoned | Unseasoned or Wet |
| Shear Plates | 0% | 20% | 33% |

### MAXIMUM PERMISSIBLE LOADS ON SHEAR PLATES

The allowable loads for all loadings except wind shall not exceed 2900 lbs for 2⅝" shear plates with ¾" bolts. The allowable wind load shall not exceed 3870#. If bolt threads bear on the shear plate, reduce the preceding values by one-ninth.

**Figure 25** By permission of TECO

**Design and Load Data for TECO Connectors**
**2 5/8" TECO Shear Plates**

# Timber Design

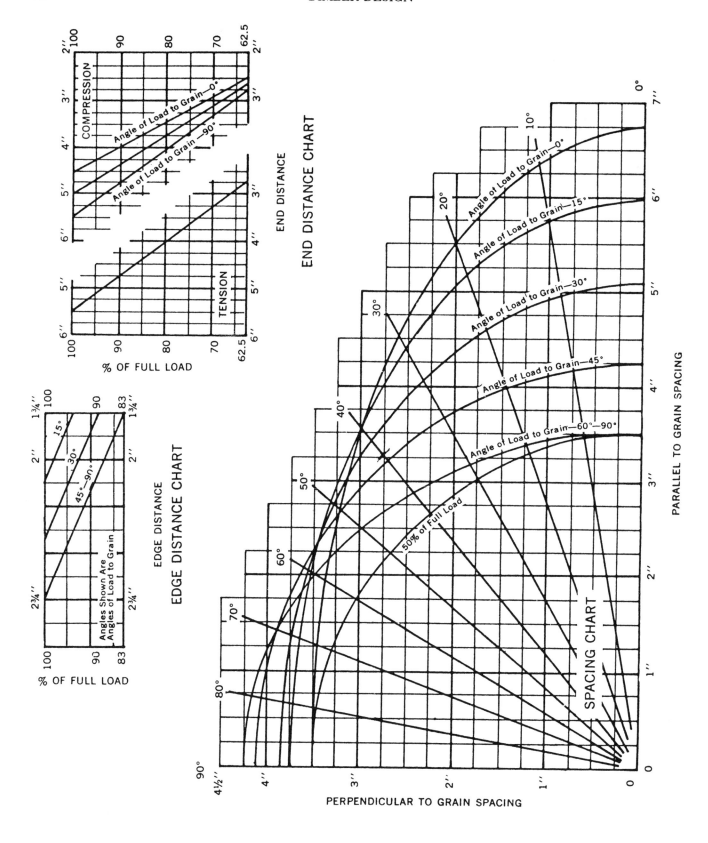

**Figure 26**    By permission of TECO
Design and Load Data for TECO Connectors
2 5/8" TECO Shear Plates

# TIMBER DESIGN

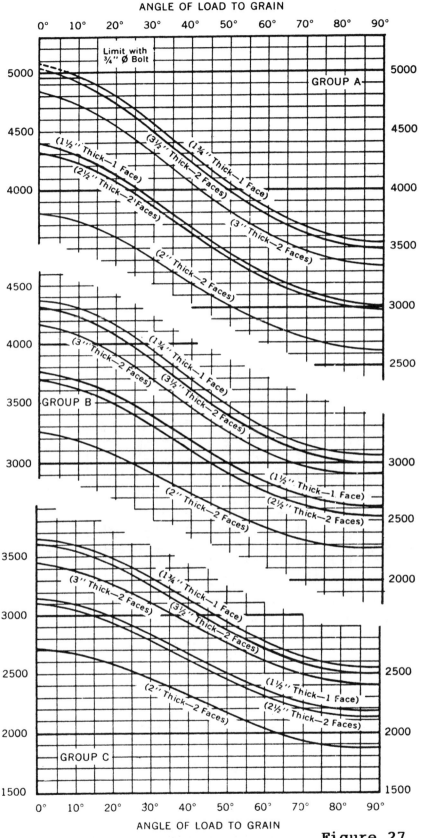

## 4" SHEAR PLATE DATA

| Shear Plates, Dimensions | Malleable Iron | Malleable Iron |
|---|---|---|
| Material | | |
| Diameter of plate | 4.03" | 4.03" |
| Diameter of bolt hole | .81" | .94" |
| Depth of plate | .64" | .64" |
| Lumber, minimum dimensions | | |
| Face, width | 5-1/2" | 5-1/2" |
| Thickness, plates in one Face only | 1½" | 1½" |
| Bolt, diameter | 3/4" | 7/8" |
| Bolt Hole, diameter in timber | 13/16" | 15/16" |
| Washers, standard, timber to timber connections only | | |
| Round, cast or malleable iron, diameter | 3 | 3-1/2" |
| Square Plate | | |
| Length of side | 3 | 3 |
| Thickness | 1/4" | 1/4" |
| (For trussed rafters and other light structures standard wrought washers may be used.) | | |
| Projected Area, for one shear plate, square inches | 2.58 | 2.58 |

## SHEAR PLATE SPECIFICATIONS

Shear plates shall be TECO shear plates as manufactured by TECO, Washington, D.C. Malleable Iron Types—Malleable iron shear plates shall be manufactured according to current A.S.T.M. Standard Specifications A 47, Grade 32510, for malleable iron castings. Each casting shall consist of a perforated round plate with a flange around the edge extending at right angles to the face of the plate and projecting from one face only, the plate portion having a central bolt hole reamed to size with an integral hub concentric to the bolt hole and extending from the same face as the flange.

## PERCENTAGES FOR DURATION OF MAXIMUM LOAD

| | |
|---|---|
| Two Months Loading, as for snow | *115% |
| Seven Days Loading | *125% |
| Wind or Earthquake Loading | *133⅓% |
| Impact Loading | *200% |
| Permanent Loading | 90% |

* Do not exceed limitations for maximum allowable loads for shear plates given elsewhere on this page.

## DECREASES FOR MOISTURE CONTENT CONDITIONS

| Condition when Fabricated | Seasoned | Unseasoned | Unseasoned |
|---|---|---|---|
| Condition when Used | Seasoned | Seasoned | Unseasoned or Wet |
| Shear Plates | 0% | 20% | 33% |

## MAXIMUM PERMISSIBLE LOADS ON SHEAR PLATES

The allowable loads for all loadings except wind shall not exceed 4970 lbs for 4" shear plates with ¾" bolts and 6760 lbs for 4" shear plates with ⅞" bolts. The allowable wind loads shall not exceed 6630 lbs when used with a ¾" bolt and 9020 lbs when used with a ⅞" bolt. If bolt threads bear on the shear plate, reduce the preceding values by one-ninth.

**Figure 27**  By permission of TECO

**Design and Load Data for TECO Connectors**

**4" TECO Shear Plates (Wood-to-Wood)**

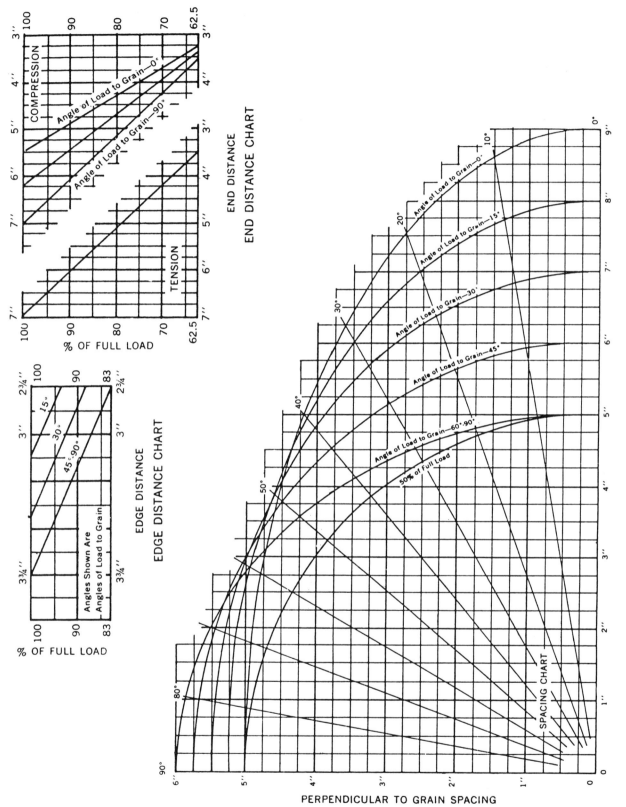

**Figure 28** By permission of TECO
Design and Load Data for TECO Connectors
4" TECO Shear Plates (Wood-to-Wood)

# TIMBER DESIGN

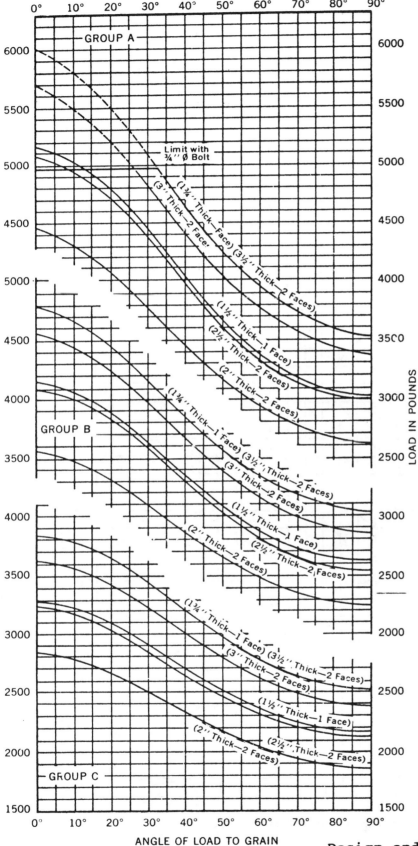

## 4" SHEAR PLATE DATA

| Shear Plates, Dimensions Material | Malleable Iron | Malleable Iron |
|---|---|---|
| Diameter of plate | 4.03" | 4.03" |
| Diameter of bolt hole | .81" | .94" |
| Depth of plate | .64" | .64" |
| Lumber, minimum dimensions | | |
| Face, width | 5-1/2" | 5-1/2" |
| Thickness, plates in one face only | 1½" | 1½" |
| Steel Shapes or Straps (Thickness required when used with shear plates) Thickness of steel side plates shall be determined in accordance with A.I.S.C. recommendations. | | |
| Hole, diameter in steel straps or shapes | 13/16" | 15/16" |
| Bolt, diameter | 3/4" | 7/8" |
| Bolt Hole, diameter in timber | 13/16" | 15/16" |
| Projected Area, for one shear plate, square inches | 2.58 | 2.58 |

## SHEAR PLATE SPECIFICATIONS

Shear plates shall be TECO shear plates as manufactured by TECO, Washington, D.C. Malleable Iron Types—Malleable iron shear plates shall be manufactured according to current A.S.T.M. Standard Specifications A 47, Grade 32510, for malleable iron castings. Each casting shall consist of a perforated round plate with a flange around the edge extending at right angles to the face of the plate and projecting from one face only, the plate portion having a central bolt hole reamed to size with an integral hub concentric to the bolt hole and extending from the same face as the flange.

## PERCENTAGES FOR DURATION OF MAXIMUM LOAD

| | |
|---|---|
| Two Months Loading, as for snow | *115% |
| Seven Days Loading | *125% |
| Wind or Earthquake Loading | *133⅓% |
| Impact Loading | *200% |
| Permanent Loading | 90% |

\* Do not exceed limitations for maximum allowable loads for shear plates given below

## DECREASES FOR MOISTURE CONTENT CONDITIONS

| Condition when Fabricated | Seasoned | Unseasoned | Unseasoned |
|---|---|---|---|
| Condition when Used | Seasoned | Seasoned | Unseasoned or Wet |
| Shear Plates | 0% | 20% | 33% |

## MAXIMUM PERMISSIBLE LOADS ON SHEAR PLATES

The allowable loads for all loadings except wind shall not exceed 4970 lbs for 4" shear plates with ¾" bolts and 6760 lbs for 4" shear plates with ⅞" bolts. The allowable wind loads shall not exceed 6630 lbs when used with a ¾" bolt and 9020 lbs when used with a ⅞" bolt. If bolt threads bear on the shear plate, reduce the preceding values by one-ninth.

By permission of TECO

**Figure 29**
Design and Load Data for TECO Connectors
4" TECO Shear Plates (Wood-to-Steel)

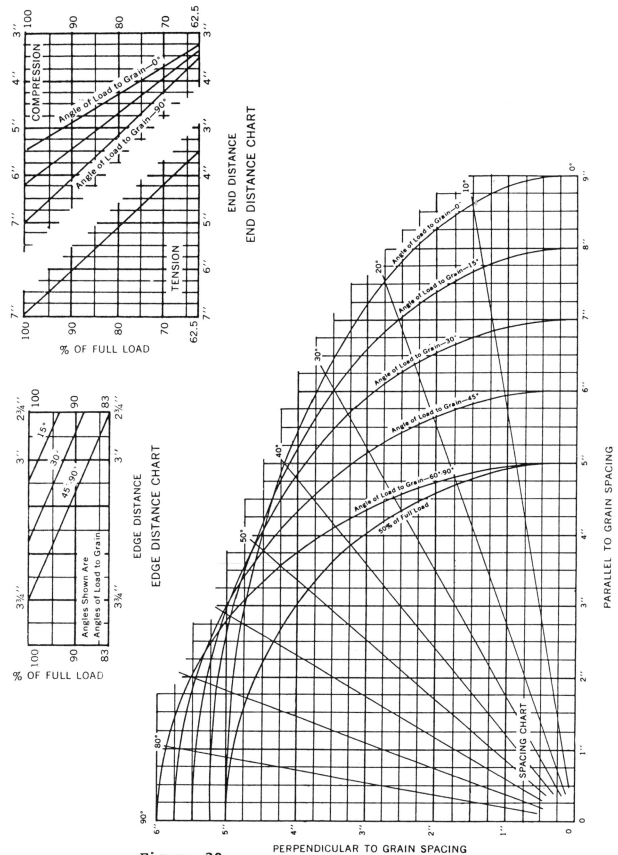

**Figure 30**
Design and Load Data for TECO Connectors
4" TECO Shear Plates (Wood-to-Steel)   By permission of TECO

**Example 11**

Check the capacity of the members and the four 2 1/2" split rings under the design snow load. Lay out the dimensions for the connectors. The connection was fabricated when dry and is sheltered from the elements.

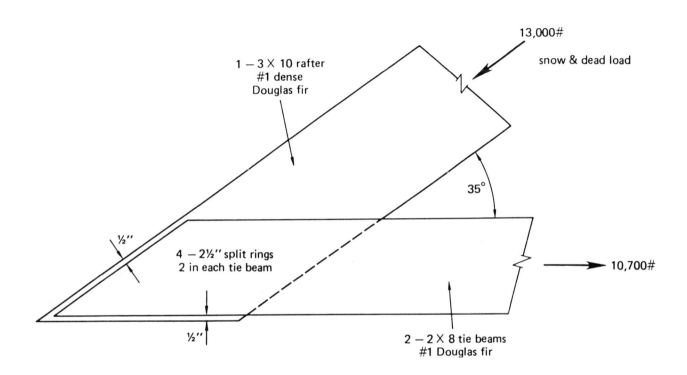

Check the net section and connector capacities in the rafter and tie beams. Laying out the connector dimensions is a complicated process with many available options. One way to systematically lay out the connection is to first consider each member separately and then to look at the entire connection. For each member, lay out the acceptable region for connectors. This region will be defined by the end and edge distances for the given connector, angle of load, and percentage of load capacity used. Then determine the acceptable connector region for the entire connection by overlaying the acceptable regions for each member. Once this region is found, various layouts for the required number of connectors can be investigated for acceptability with the connector spacing chart.

First, find the net section of the 3x10 rafter.

$(23.125)$ in$^2$ − $(2)(1.10)$ in$^2$ − $(9/16")[2.5" − (2)(.375")] = 19.94$ in$^2$. Appendix 4 has the 1.10 in$^2$ connector area and the .375" connector depth. The capacity is

$(19.94)$ in$^2$ $(1.15)(7/8)(1450)$ psi = 29,100 lb

The 1.15 load duration factor comes from table 2. The 7/8 is an allowance for knots in the net section.

The connector capacity in the rafters, with four 2 1/2" split rings, a group A species, 35° to grain, and the 2" THICK-2 FACES curve, is 2830 lb/connector (see figure 21). The rafter split rings capacity is

$(2830)$ lb/connector $(4)$ connectors $(1.15)$ = 13,000 lb

The connectors are adequate, and at 100% of capacity.

Now the tie beams' capacity is checked. The net section of the 2-2x8 is

$(2)[(10.875)$ in$^2$ − $(1.10)$ in$^2$ − $(9/16")(1.5" − 0.375")] = 18.28$ in$^2$, where the connector dimensions are found in appendix 4. The tie beam capacity is

$(18.28)$ in$^2$ $(1.15)(7/8)(1000)$ psi $(.8)$ = 14,700 lb

The connector capacity in the tie beams is found in figure 21 for 2 1/2" split rings, load at 0° to grain, a group B species, and the 1 1/2" THICK-1 FACE curve to be 2740 lb/connector. The capacity of the four connectors is

$(2740)$ lb $(4)$ connectors $(1.15)$ = 12,600

This is also greater than the applied 10,700 lb load.

The connection has adequate capacity, as long as the connectors fit into the available space without violating minimum spacing requirements. These spacing requirements should be checked next.

First, lay out the acceptable area for connectors in the rafter, as limited by edge and end distances. This member is in compression and at 100% capacity.

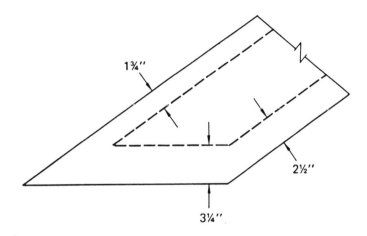

Next, the tie beams can have connectors in this region, considering that these connectors are at (10,700) lb/(12,600) lb, or 85% of their capacity and in tension.

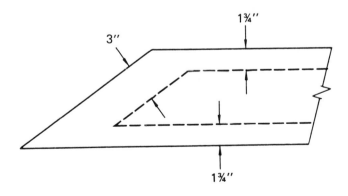

Overlapping these two regions, while maintaining the half-inch clearances, means the connectors can be located anywhere in the common region. The more restrictive of any two parallel limits is the one which controls.

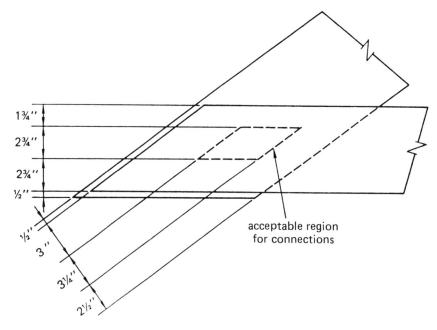

acceptable region for connections

Within the acceptable region, many options remain for spacing the connectors. The widest allowable spacing should be checked first, to make sure there is even room for the required connectors to fit and still meet the spacing requirements between individual connectors. If there is extra room, the engineer should compromise between the widest spacing permitted by the edge and end distances, and the closest spacing allowed for the given connectors. The spacings must be checked parallel and perpendicular to the grain in each of the connected members.

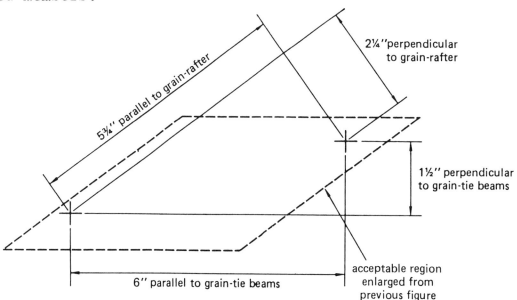

This set of connector spacings fulfills all the spacing and clearance requirements in each connected member. It also contains at least one set of reasonably simple layout dimensions.

# Example 12

Check the 2 5/8" shear plates for capacity and determine the spacing requirements for their layout. Determine the minimum required lag bolt size. The connection will be kept dry in service.

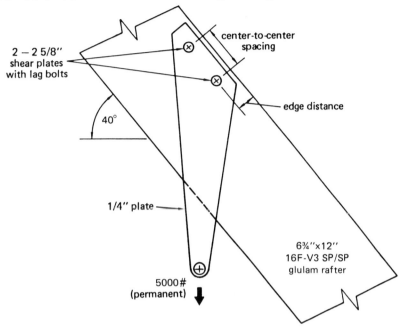

The connector capacity is found in figure 25. The lag screw used to hold the connector in place must be considered. The angle of load to grain has to be determined. The spacing considerations will include edge distances and center-to-center spacings.

The glulam part of appendix 2 gives a $F_{c\perp}$ of 650 psi for this glulam member. Footnote 14 to the same appendix then puts the connectors in a group A (or II) species. The connector capacity is found in figure 25 with the load at 40° to the grain and the 2½" THICK-2 FACES curve, to be 2820 lb/connector. The 2 FACES curve was used even with a connector in only one face because it is conservative to assume a one-face connector is as strong as if there were another connector in the far face.

This means the connection capacity is

(2) connectors (2820) lb/connector (.9)(1.0)(1.0) = 5,075 lb.

The 1.0 is the lag bolt modification factor, found in table 14. The other 1.0 is the CUF of table 4, and the .9 is the LDF from table 2.

The connection is adequately strong. Now, establish the required spacings.

The connectors are loaded at (5000) lb /(5075) lb, or 98.5% of their capacity. Figure 26 contains the necessary charts. The edge distance is an unloaded one, and must be at least 1 3/4". The center-to-center spacing must be at least 4 1/2". This parallel with the grain spacing was interpolated for 40° load angle and 98.5% of capacity, with the perpendicular with the grain spacing equal to zero.

The lag screw, in a group II species, is determined from table 14 to require a 3 1/2" diameter's penetration. With a 3/4" lag screw, this means a (3.5)(.75") + 0.25" = 2.875" minimum length. The .25" is included to account for the metal plate. Therefore, use a 3/4"x3" lag screw.

Note the difference in meaning for "lag bolt penetration." In lateral loading, the length of lag screw in the timber is the penetration. With withdrawal loads, the penetration is the length of the threaded portion of lag screw in the member.

## 7. REVIEW OF ESSENTIAL MECHANICS OF MATERIALS CONCEPTS

### A. Introduction

The material presented in this section is not intended as a complete treatment of Mechanics of Materials. Rather, it stresses the specifics and modifications of classic Mechanics of Materials concepts as they apply to timber structural elements. The section is organized by loads in the axial and transverse directions, and their combinations.

### B. Axial Loadings

Axial loads are applied along the longitudinal axis of the member. In wood, this means the loads are oriented with the fibers, or in the strong direction. Axial loads can be either tension or compression. If the load were aligned perfectly with the centroid of the cross section, the stresses caused by axial loads would be uniform on a cross section, at some distance from the ends.

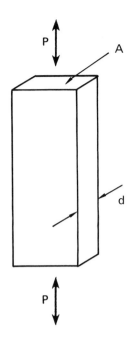

**Figure 31
Member with Axial Load**

Buckling

When an axially loaded member, commonly called a column, is in compression, a phenomenon known as buckling can drastically reduce the load capacity of the member. Buckling can be a reversible bend that develops in the middle of a column, or it can be a sudden and drastic collapse initiated by lateral movement.

The basics of column buckling were set down by Leonhard Euler in 1744. The Euler equation is a solution for the stress level at which a given column becomes unstable. At stress levels higher than this critical stress, any lateral displacement causes increased bending stresses and column failure.

$$\text{Euler's critical stress} = \frac{\pi^2 E}{(L/k)^2} \tag{4}$$

With a rectangular cross section: b X d; (b > d), common in wood columns, the Euler equation becomes:

$$f_{cr} = \frac{\pi^2 E}{12(L/d)^2} \tag{5}$$

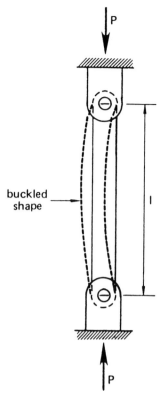

**Figure 32
Column Subject to Buckling**

The quantity L/d is known as the slenderness ratio of a given column. With wood columns, a factor of 2.74 is included to account for material variability, safety, and a correction of tabled E values. This results in the following expression for allowable stresses in timber columns:

$$F_c' = \frac{0.3E}{(L/d)^2} \tag{6}$$

Limitations in material strength place upper bounds on the applicability of the Euler equation, and wood is no exception. Figure 33 is a graph showing column failure stresses as a function of their slenderness ratios.

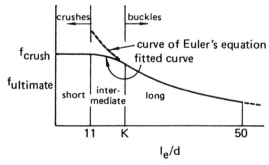

**Figure 33
Column Failure Stresses**

# Timber Design

The National Design Specifications divide the range of slenderness ratios into three regions:

$L/d \leq 11$: Short Column

$11 < L/d < K$; Where $K = 0.671 (E/F_c)^{0.5}$: Intermediate Column

$K \leq L/d < 50$: Long Column

- **Effective Column Length $L_e$:** The Euler equation and the rest of the design equations are based on pinned-pinned end conditions. The design formula can be used with other end conditions by modifying the column length to an effective length, $L_e = K_e L$. The most common end conditions and their effective lengths are included in table 16.

## Axial Deflections

The longitudinal deformation for both tension and compression members is

$$\text{Change in Length} = PL/AE \tag{7}$$

The tabulated E value in appendix 2 is an average value that has been reduced in order to adjust bending deflections to include shear. The "true E" is about 9% larger than the table value, but this is still an average value.

**Table 16**
**Effective Column Lengths**
$$L_{effective} = K_e L$$

| Buckling modes | | | | | | |
|---|---|---|---|---|---|---|
| Theoretical $K_e$ value | 0.5 | 0.7 | 1.0 | 1.0 | 2.0 | 2.0 |
| Recommended design $K_e$ when ideal conditions approximated | 0.65 | 0.80 | 1.2 | 1.0 | 2.10 | 2.4 |
| End condition code | | Rotation fixed, translation fixed | | | | |
| | | Rotation free, translation fixed | | | | |
| | | Rotation fixed, translation free | | | | |
| | | Rotation free, translation free | | | | |

## C. Transverse Loading

Transverse loads are applied perpendicular to the longitudinal axis of the member. Transverse loading causes two major stress types in the member or beam. These distinct stresses are longitudinal bending stresses and shear stresses.

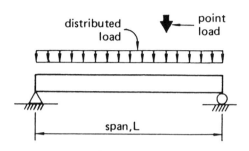

**Figure 34
Transversely Loaded Member**

### Bending Stresses

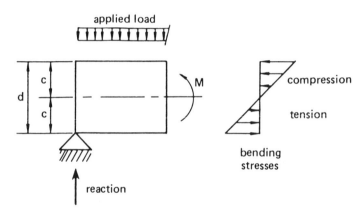

**Figure 35
Longitudinal Bending Stresses**

The classic equation for maximum bending stress at a cross section is

$$f_b = Mc/I = M/S \qquad (8)$$

Wood significantly violates several of the assumptions upon which this equation is based. To be able to use this common equation, the tabulated allowable bending stress values are actually Moduli of Rupture.

### Shear Stresses

The standard expression for shear stresses anywhere in a cross section is

$$f_v = VQ/Ib \qquad (9)$$

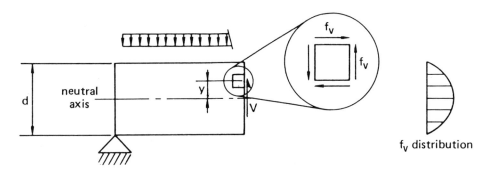

**Figure 36**
Shear Stresses due to Transverse Loads

In the rectangular cross sections common to wood members, the maximum shear stress is at the neutral axis where $Q = bd^2/8$ and $I = bd^3/12$, giving a maximum shear stress of

$$f_v = 3V/2bd = 1.5V/A \tag{10}$$

There are shear stresses in two directions -- along and across the grain. The shear strength across the grain is so much higher as to be of almost no concern. Since most transversely loaded members are horizontal beams, the shear stress component along the grain is horizontal. This is the only reason the tables give allowable "horizontal shear stresses."

### Deflections in Transversely Loaded Members

There are many numerical methods and copious tables available to determine bending deflections. In most materials, deflections due to bending stresses are so much greater than those due to shear stresses that the shear component is neglected. This is not true in timber, which has a relatively low shear modulus of elasticity.

There are two ways to deal with the shear deflections. The standard method is to ignore it and use a reduced modulus of elasticity in calculating bending deflections, which are therefore increased in compensation. This reduced E is the value found in the design tables.

The other method is to evaluate the two components separately and add them. A reasonable value for the shear modulus is 1/16 of the modulus of elasticity. In calculating the bending deflection, the true modulus must be used to avoid providing for shear deflections twice. The true modulus of elasticity can be taken as 9% higher than the tabulated value.

## Lateral Stability of Bending Members

The compression zone of any bending member acts as a column, and is similarly subject to buckling. When designing beams, the engineer must ensure that the member will either be stiff enough to resist lateral buckling itself, or be adequately restrained against sidesway. The other option is to reduce the allowable compression stresses in the member to a level safely below those which could cause buckling. This reduced bending stress, $F_b'$, is discussed in the beam design section of this book.

### D. Combined Loadings

When a member is subjected to both axial and transverse loads, the longitudinal bending stresses and axial stresses are superimposed. Since timber beams do not actually behave as the bending stress equation assumes, the two longitudinal stress types are treated independently in an interaction equation, rather than being compared with some allowable value. The interaction equation includes a buckling consideration.

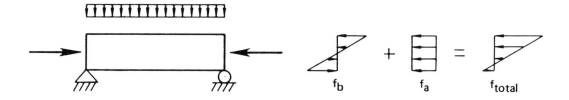

**Figure 37**
**Member Subjected to Combined Loading**

## Transverse and Axial Tension Loading

There are two interaction equations used with combined transverse and axial tension loading -- one for the tension region where the two stresses add, and the other for the compression region where the net value is considered in light of potential buckling.

In the tension zone:

$$\frac{f_t}{F_t} + \frac{f_b}{F_t} \leq 1.0 \tag{11}$$

In the compression zone:

$$\frac{f_b - f_t}{F_b'} \leq 1.0 \tag{12}$$

## Transverse and Axial Compression Loading

Buckling is a real concern in these load cases, because the transverse load tends to displace the member laterally, the very definition of buckling. Just as with simple columns, members subjected to this combined loading are divided into categories by their slenderness ratio.

The basic interaction equation is:

$$\frac{f_c}{F_c'} + \frac{f_b}{F_b' - Jf_c} \leq 1.0 \tag{13}$$

The J factor includes the buckling consideration in these load cases.

$$0 \leq J = \frac{(L_e/d) - 11}{K - 11} \leq 1.0 \tag{14}$$

Note that with $L_e/d < 11$, short columns, $J = 0$. With $L_e/d \geq K$, long columns, $J = 1.0$. It is only the intermediate columns which are actually affected by J.

## Eccentric Loading

An eccentric load is parallel to the longitudinal axis of the loaded member, but displaced from the centroid of the section. The displacement, or eccentricity, of the load causes a bending moment whose stresses must be combined with the axial stresses. Appendix H of the National Design Specifications contains many simplified design equations for various eccentric load cases.

$$\frac{f_c}{F_c'} + \frac{f_b + f_c(6 + 1.5J)(e/d)}{F_b' - Jf_c} \geq 1.0 \qquad (15)$$

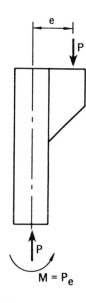

**Figure 38**
Column with Eccentric Load

## 8. DESIGN OF TIMBER STRUCTURAL MEMBERS -- AXIAL MEMBERS

Members subjected to simple axial loads include truss members, posts and columns, and diaphragm chords. Tension members are designed by simply comparing actual stresses at critical sections with allowable stresses. In compression members, stability considerations make designing more complex.

### A. Tension Members

The design process for tension members is a simple comparison of stresses at critical sections with the allowable stresses. The allowable stress is modified according to the design conditions. The design equation is:

$$f_t = P/A \leq F_t' \tag{16}$$

The load duration and condition of use factors apply to $F_t'$. Footnote 3 of the design tables contains reductions as high as 40% in the larger sizes. Net sections are reduced by 1/8 to account for the possibility of a knot being in the critical section. Design examples for tension members can be found in several of the connection design examples.

### B. Compression Members

There are three types of timber columns: solid, spaced, and built-up. Solid columns are the simplest and most common, and they will be used to illustrate the basic design procedures for all columns. Spaced columns use timber more efficiently, but they are costlier. Spaced columns are generally only found in trusses where they are required due to layout considerations. Built-up columns are solid assemblies of smaller pieces. A capacity reduction is necessary with built-up columns because of the imperfect connection between the separate pieces.

The basic design equation for compression members looks as simple as the tension member equation:

$$f_c = P/A \leq F_c' \tag{17}$$

The cross sectional area is either net or gross because there are two distinct concerns in compression members: crushing at connections, and buckling in the main body of the member. Depending on where the column is braced relative to connections, one of the areas or the other will be used.

## Connection at a Braced Point

If the connection is at a point which is restrained against buckling, both areas are used.

Gross area is used to calculate the stress which is compared to the allowable buckling stress, $F_c'$.

Net area is used to calculate the stress at the connection for comparison with $F_c$, the allowable crushing stress.

## Connection at an Unbraced Point

If the connection is at a point which is not restrained against buckling, only the net area is used to calculate the actual stress level. This actual stress level should be compared with the lower allowable buckling stress, $F_c'$.

## $F_c'$, Allowable Compression Stress, Considering Stability

The major factor determining allowable buckling stress is a member's slenderness ratio, $L_e/d$. Both elements of the slenderness ratio will vary with column configuration.

- **Determining $L_e$, Effective Column Length:** The Euler buckling stress equation is based on an assumed pinned-pinned column. While this is the most common end condition in timber columns, it is possible to analyze other conditions with the same equations by substituting an effective length for the column. To find a column's effective length, multiply its actual length by the appropriate factor from table 16.

  There may be more than one effective length to consider in a given column. (See the following section for more information.) A fixed end condition is very difficult to achieve in timber, and it is generally not conservative to assume one exists.

- Determining d, "Minimum" Column Cross Sectional Dimension: The cross sectional dimension, d, which is used to determine a column's slenderness ratio, will vary with bracing layout.

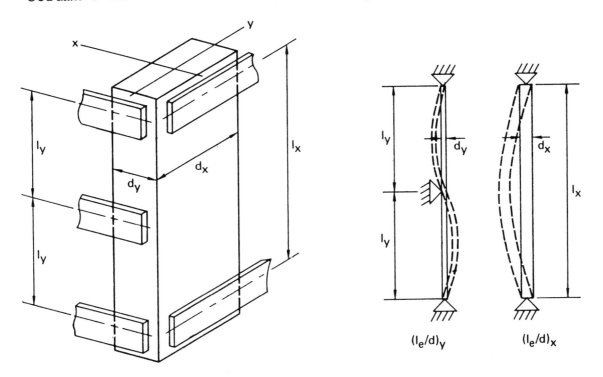

**Figure 39**
**Determining the Slenderness Ratio for a Column**

It is important to match the length of column between bracing points with the cross sectional dimension associated with buckling in the same plane as the braces. In certain cases, both ratios must be calculated, with the larger slenderness ratio controlling.

In any case, the dimension d is always an actual dimension, not a nominal one.

Column Slenderness Categories

The maximum slenderness ratio determines the length category for any column. There are three of these categories, each with its own allowable stress.

- Short Columns: If $(L_e/d)_{max} \leq 11$, the column is short. It will crush before it buckles. Therefore, there is no reduction in the allowable compression stress parallel to the grain.  Equation 18 applies.

$$F_c' = F_c \tag{18}$$

- **Intermediate Columns:** If $(L_e/d)_{max} > 11$, but still less than K, the column is an intermediate column which may fail in crushing, buckling, or both.  The upper limit of this length range, K, is a function of the material properties.

$$K = 0.671(E/F_c)^{0.5} \tag{19}$$

The load duration factors of table 2 are applied to $F_c$ in evaluating K for a specific instance, but they are not applied to E.  Any applicable condition of use factors are applied to both E and $F_c$.  The tabulated values for E are average values, meaning that there is a 50% chance of a column having a lower modulus of elasticity than the table value.  For critical columns, the conservative approach will be to use one-half the tabulated E values to calculate capacities.

The allowable stress in an intermediate column is

$$F_c' = F_c\left[1 - \frac{1}{3}\left(\frac{L_e/d}{K}\right)^4\right] \tag{20}$$

The load duration factor is also applied to $F_c$ in this equation.  Note that this means the load duration factor figures in both K and $F_c$ of equation 20.

- **Long Columns:** If $(L_e/d)_{max} \geq K$, the column is a long column and is subject to buckling.  The Euler equation is used to determine the allowable stress:

$$F_c' = 0.30E/(L_e/d)^2 \tag{21}$$

Note that since the load duration factor does not apply to E, the duration of a load has no effect on the allowable stress for a long column. For critical columns, one-half of the tabulated E value should be used.

The upper allowable limit on the slenderness ratio for a solid column is 50. Columns more slender than this limit are too slender and should be braced or increased in size.

# Example 13

Design a 12 foot long, #1 Douglas fir-larch column to carry a combined snow and dead load of 14 kips. The column will be pinned-pinned at its ends and braced in one direction at its centerline. Assume the column is dry when installed, and is protected from the elements.

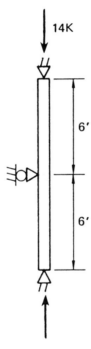

Since the allowable stress in a column is a function of its slenderness ratio, column design is necessarily a trial and error process.

In order to pick a reasonable first trial size, estimate an allowable stress. Assuming dimension lumber sizes, appendix 2 gives an $F_c$ of 1250 psi and an E of 1,800,000 psi. The LDF of table 2 is 1.15 for snow loads. If the load duration and buckling considerations were to offset each other, the required cross sectional area would be equal to (14,000) lb/(1250) psi = 11.2 in$^2$.

Look in appendix 5 for a section with this area. The cross sectional dimension ratio should be about 2:1 in order to make efficient use of the available bracing. Select a 3x6 which has an area of 13.750 in$^2$. Now, evaluate the slenderness ratios for the 3x6 trial size.

$(L_e/d)_{weak} = (6)$ ft $(12)$ in/ft $/(2.5)$ in $= 28.80$

$(L_e/d)_{strong} = (12)$ ft $(12)$ in/ft $/(5.5)$ in $= 26.18$

The two slenderness ratios are nearly equal, which is efficient, but weak axis buckling controls. The maximum slenderness ratio is greater than 11, so the column is not short. To distinguish between long and intermediate columns, K must be evaluated.

$K = 0.671(E/F_c)^{0.5} = 0.671[(1,800,000)$ psi $/(1.15)(1250)$ psi$]^{0.5}$
$= 23.74$

Note that the 1.15 LDF appears in the expression for K. Since the maximum slenderness ratio is greater than K, the column is long. From equation 21,

$F_c' = (.3)(1,800,000)$ psi $/(28.80)^2 = 651$ psi

The 3x6 capacity is

$(651)$ psi $(13.75)$ in$^2 = 8,950$ lb $= 8.95$ kips.

This capacity is much lower than the required 14 kips, so try the next larger size with the 2:1 dimension ratio, a 4x8.

$(L_e/d)_{weak} = (6)$ ft $(12)$ in/ft $/(3.5)$ in $= 20.57$
$(L_e/d)_{strong} = (12)$ ft $(12)$ in/ft$/(7.25)$ in $= 19.86$

The weak axis buckling still controls. K does not change, but it is larger than the maximum slenderness ratio, making this an intermediate column. From equation 20,

$F_c' = (1.15)(1250)$ psi $[1 - (1/3)(20.57/23.74)^4] = 1167$ psi

## Timber Design

The 4x8 column capacity is

$$(1167) \text{ psi } (25.375) \text{ in}^2 = 29{,}600 \text{ lb.}$$

This section is adequate, but significantly oversized. Try a 4x6.

$$(L_e/d)_{weak} = 20.57 \text{ (same as for a 4x8)}$$
$$(L_e/d)_{strong} = (12) \text{ ft } (12) \text{ in/ft } / (5.5) \text{ in} = 26.18$$

The 26.18 slenderness ratio is greater than K (23.74), so the column is long. From equation 21,

$$F_c' = (.3)(1{,}800{,}000) \text{ psi } / (26.18)^2 = 788 \text{ psi}$$

The 4x6 capacity is

$$(788) \text{ psi } (19.25) \text{ in}^2 = 15{,}169 \text{ lb} = 15.2 \text{ kips.}$$

This is greater than the required 14 kips, but not enough to warrant further investigation of smaller sections.

Use a 4x6, #1 Douglas fir-larch column, with the midspan bracing in the weak direction.

## Example 14

(a) A column is a 10x10, dense #1 Douglas fir-larch, installed wet and exposed in service. The member is 25 feet long, with the bottom 6 feet embedded and the top braced against sidesway. What is the dead load capacity of this column, assuming lives depend upon its not failing? (b) Find the minimum required bearing area at the top, with and without a metal bearing plate, to distribute the load. (c) What would the capacity be if only 5 feet were embedded?

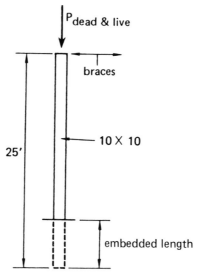

(a) To find the capacity, establish the $L_e/d$ ratio using the effective length. Check the end grain bearing stress to establish the required bearing areas.

From table 16, the exposed 19 feet of column has an effective length of

$$L_e = (19) \text{ ft } (0.8) = 15.20 \text{ ft}$$

The 0.8 factor accounts for the fixed bottom, as seen in table 16. The slenderness ratio is, therefore,

$$(15.20) \text{ ft } (12) \text{ in/ft } /(9.5) \text{ in} = 19.20$$

The division between intermediate and long columns in dense #1 Douglas fir-larch is found from values in appendix 2.

$$K = (0.671)[(1.0)(1,700,000/2) \text{ psi}/(0.91)(0.9)(1200) \text{ psi}]^{.5}$$
$$= 19.73$$

From table 2, a load duration factor for dead load is applied to $F_b$. The modulus of elasticity was halved because this is a critical column. The 0.91 and 1.0 condition of use factors come from footnote 10 of appendix 2 and account for the wet use. Since the slenderness ratio is less than K, the post acts as an intermediate column. From equation 20,

$$F_c' = (0.9)(1200)(.91) \text{ psi } [1 - 1/3(19.20/19.73)^4] = 689 \text{ psi}$$

This allowable stress level gives a post capacity of

$$(689) \text{ psi } (90.25) \text{ in}^2 = 62{,}200 \text{ lb}$$

(b) The allowable end grain bearing stress for dense Douglas fir-larch is 1570 psi, from appendix 1 assuming wet service conditions. Modifying for dead load duration with the 0.9 of table 2, the minimum required bearing area required is

$$(62{,}200) \text{ lb}/(0.9)(1570) \text{ psi} = 44.0 \text{ in}^2$$

Note that this area is based on end grain stresses at 100% of capacity, which requires a metal bearing plate to distribute the stress. The plate can be omitted by holding stress levels to only 75% of the allowable stress. This would require a minimum net bearing area of

$$(62{,}200) \text{ lb}/(0.9)(1570) \text{ psi }(.75) = 58.7 \text{ in}^2$$

(c) If the column was embedded only 5 feet, the slenderness ratio would be

$$L_e/d = (20) \text{ ft }(12) \text{ in/ft }(0.8)/(9.5) \text{ in} = 20.21$$

This is just larger than K, making the post a long column. The allowable stress is, therefore,

$$F_c' = [(.3)(1{,}700{,}000)/(2) \text{ psi}]/(20.21)^2 = 624 \text{ psi}$$

The post capacity is

$$(624) \text{ psi }(90.25) \text{ in}^2 = 56{,}300 \text{ lb}.$$

### Spaced Columns

A spaced column consists of two or more solid columns connected to one another, but separated along their length by spacer blocks. The separated members act together much as an I-beam section does, gaining considerable bending stiffness in what would otherwise be the weak plane for buckling. The internal lateral bracing makes the spaced column a more efficient member. This construction method is significantly more costly, and it is used only when the configuration is also justified by fabrication considerations.

In order to resist the bending shear deformations that would allow the spaced column to buckle, the connections must be stiff. Determining the spacer block connection stiffness and evaluating a variety of slenderness ratios are the essential elements of spaced column design. The National Design Specifications section 3.8 contains the necessary particulars, should a spaced column seem worth the effort and expense.

## Built-up Columns

It is sometimes easier to build up a large column from smaller sections than to find a large or dry enough solid-sawn member. Unlike spaced columns which are connected rigidly enough to achieve full composite member action, built-up columns require reduced capacities to account for the non-organic connection between the individual pieces. Table 17 gives these reductions as they vary with the slenderness ratio of the built-up section.

**Table 17**
**Percent Reductions in Strength for Built-Up Columns**

| L/d Ratio | Percent Reduction |
|---|---|
| 6 | 18 |
| 10 | 23 |
| 14 | 29 |
| 18 | 35 |
| 22 | 26 |
| 26 | 18 |

# 9. DESIGN OF TIMBER STRUCTURAL MEMBERS -- BENDING MEMBERS

Timber bending members are designed with the allowable stress method. Four separate concerns are dealt with in this chapter: bending stresses, shear stresses, deflections, and bearing stresses. Differences in designing solid sawn and glue-laminated members will be covered as they arise.

### A. Bending Stresses

Actual bending stresses are evaluated with equation 22, and they should be less than the adjusted allowable bending stress.

$$f_b = Mc/I = M/S \leq F_b' \tag{22}$$

The bending moment, M, is determined in the normal manner. The section modulus, S, is found in appendix 5 for nominal lumber sizes. The section modulus for the net section must be calculated at connections and notches.

There are two ways that a beam fails from high bending stresses. A beam can either buckle laterally, or the extreme fibers can fail. Since these are two distinct phenomena, there are two separate allowable bending stresses to be evaluated, with the lesser controlling.

#### Allowable Bending Stresses, Lateral Buckling

For anything other than clearly well-braced cases, the allowable bending stress considering lateral buckling is a function of the beam's slenderness. If a beam is no deeper than its thickness, or if the beam is supported full-length along its compression edge, lateral buckling is ignored. Otherwise, the beam slenderness factor, $C_s$, must be calculated.

$$C_s = (L_e d/b^2)^{.5} \tag{23}$$

The d and b dimensions are specific to the beam cross section, while the effective length, $L_e$, is a function of an unbraced length, $L_u$, and the beam/loading configuration.

The effective beam length, $L_e$, is a multiple of the unbraced length, $L_u$. The unbraced length is the distance between lateral supports or points where rotation is prevented. The effective length is obtained by multiplying the unbraced length by the appropriate factor of figure 40.

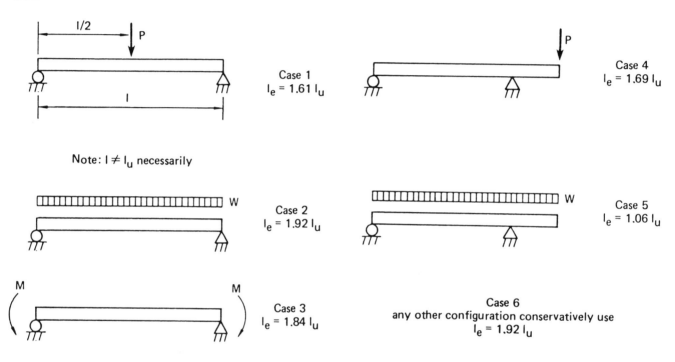

**Figure 40
Effective Beam Lengths**

The effective length factors of figure 40 are based on an assumed $L_u/d$ ratio of 17. For other ratios, the $L_e$ found with figure 40 can be multiplied by: $0.85 + 2.55/(L_u/d)$. This additional adjustment of $L_e$ does not apply to cases 2, 3, and 6 of figure 40.

If a single beam, continuous over supports, has more than one slenderness factor along its length, the maximum one controls.

## Beam Slenderness Categories

Once the slenderness factor, $C_s$, is determined, the beam can be classified as short, intermediate, or long. Each beam category has its own allowable stress levels.

- **Short Beams:** If $C_s \leq 10$, the beam is short and $F_b' = F_b$.

- **Intermediate Beams:** If $10 < C_s \leq C_k = (0.811)(E/F_b)^{.5}$, the beam is intermediate.

$$F_b' = F_b[1 - (1/3)(C_s/C_k)^4] \tag{24}$$

- **Long Beams:** If $C_k < C_s$ and no greater than the upper limit of 50, the beam is a long one.

$$F_b' = (0.438)(E)/(C_s)^2 \tag{25}$$

There are conflicting opinions about the application of load duration and condition of use factors in these equations. The author suggests applying the load duration factor to $F_b$ and the condition of use factor to E and $F_b$ wherever possible. This is a fairly standard approach.

## Allowable Bending Stresses, Limited by Extreme Fiber Failure

There are several cumulative adjustments that apply to the allowable bending stress as limited by fiber failure.

- **Load Duration:** The load duration factors of table 2 apply to $F_b$.

- **Conditions of Use:** The footnotes in appendix 2 contain modifications for various conditions of use.

- **Form Factor:** The bending stress design equation assumes a rectangular timber section. With a circular cross section, $F_b$ is multiplied by 1.18. If a rectangular section is used on edge, as a "diamond", $F_b$ is multiplied by 1.414, regardless of which direction the load is applied.

- Size Factor: As timber beams deepen, they behave less in accordance with the bending stress equation assumptions. The modification made to $F_b$ to account for this difference, $C_F$, is the only factor not applied in lateral stability considerations. Calculation of $C_F$ depends on whether the beam is solid-sawn or glue-laminated.

In solid sawn beams deeper than 12 inches, $F_b$ is multiplied by the size factor, $C_F$.

$$C_F = (12/d)^{1/9} \tag{26}$$

The size factor, $C_F$, used with glue-laminated beams, is more complex, varying with depth and load type. See table 18 for these factors.

### Table 18
### Size Factors for Glue-Laminated Beams

| depth d (in) | uniformly distributed load | single concentrated load | third-point loading (2 equal loads at L/3 and 2L/3) |
|---|---|---|---|
| 12 | 1.00 | 1.08 | .97 |
| 19 | .95 | 1.02 | .92 |
| 31 | .90 | .97 | .87 |
| 52 | .85 | .92 | .82 |
| 90 | .80 | .86 | .77 |

The table 18 factors may be interpolated for intermediate beam depths. The third point loading refers to two loads evenly spaced along the span.

## Curved Bending Members

There are two additional concerns with glue-laminated beams fabricated with an in-plane curvature larger than simple camber.

- **Curvature Factor:** For only those regions which contain bent laminae, the allowable bending stress is multiplied by the curvature factor, $C_c$.

$$C_c = 1 - 2{,}000(t/R)^2 \qquad (27)$$

The radius, $R$, is measured in inches to the member inside face.

The $t/R$ ratio is limited by splitting to $1/100$ for Southern pine, and $1/125$ for other softwoods.

- **Radial Tension or Compression:** Bending moments which tend to straighten a bent member cause radial tension. Bending moments that tend to increase the curvature cause radial compression. In either case, the radial stress in a member with constant cross section is

$$f_r = 3M/2Rbd \qquad (28)$$

The radius, $R$, is measured in inches to the member centerline.

For curved beams with varying cross sections, see the National Design Specifications.

The allowable radial tension, $F_{rt}$, is 1/3 the $F_v$ for Southern pine and Douglas fir-larch with wind and seismic loads. For other load types in Douglas fir-larch, $F_{rt}$ is only 15 psi. These stress levels are so low that mechanical radial reinforcing, such as bolts or lag screws, may be required.

The allowable radial compression, $F_{rc}$, in glue-laminated beams varies with the species and grade, as seen in table 19.

### B. Shear Stresses in Bending Members

The design equation for shear stresses in rectangular section beams is

$$f_v = 1.5V/A \leq F_v' \qquad (29)$$

The tabulated allowable shear stress for sawn lumber, $F_v$, is conservatively based on an assumed worst-case split at the member's end. The National Design Specifications recognizes how restrictive this assumption can be and essentially allows the tabulated $F_v$ to be increased by 50%, if the maximum possible shear force is carefully evaluated. This increase is not applicable to glue laminated members.

**Table 19**
**Allowable Radial Compression -- Curved Glue-Laminated Beams**

| Species | Combination Symbol | $F_{rc}$ (psi) |
|---|---|---|
| Douglas fir-larch | All | 385 |
| Southern pine | 16F-1, 16-F2, 20F-3, 20F-4, 24F-4, 6, 7 | 255 |
| Southern pine | All others | 385 |

There are several special circumstances in shear calculations which deserve mention.

<u>Shear With Loads Near Supports:</u> Most materials used in bending members are stiff enough across the member depth that loads very near the supports are transferred directly to the support by arching compression action, rather than through bending. Shear is critical enough in timber design that it is worth neglecting these loads. Therefore, any and all loads applied closer to the support than the beam depth may be neglected.

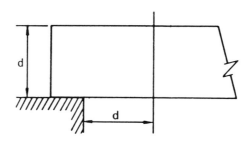

**Figure 41**
**Loads Near Supports**

<u>Shear in Notched Beams:</u> It is tempting to solve many clearance problems by notching timber beams. Unfortunately, this not only reduces the cross section available to resist shear, but it also produces serious stress concentrations. Therefore, shear stresses at a notch are calculated by including a factor which reflects the severity of the notching.

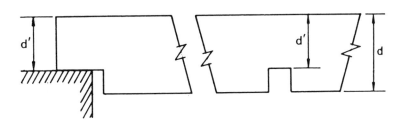

**Figure 42**
**Shear Stresses in Notched Beams**

$$f_v = \frac{3}{2} \frac{V}{bd'} \frac{d}{d'} \qquad (30)$$

The stress concentration factor, $d/d'$, can be neglected if the notch has a gradual transition. Notched beams are a common source of problems, and should be used only if there is no other reasonable option.

Shear in Connections: If a connection is more than 5 member depths from the end of the member, splits and their effects are reduced. Therefore, tabulated allowable shear stresses can be increased 50% at such connections. However, the shear stress must be evaluated on the basis of the effective depth of the member at the connection. This is often equal to the unloaded edge distance, but the National Design Specifications should be consulted in questionable cases.

Shear in Checked Beams: Large splits in wood are most often found at the ends of members, where drying defects can propagate. This is also where shear stresses are highest in simple beams. The combination of the worst defect appearing at the region of highest corresponding stress can be alarming when viewed from below. Even though the beam must act as two separate members stacked on top of each other, it is stronger than calculations might show because the shear stress in the two pieces is redistributed nearly equally throughout the cross section. See appendix E of the National Design Specifications for more information.

## C. Beam Deflections

The displacements of timber beams under loads are calculated almost exactly as they are with other structural materials. The modulus of elasticity is given in appendix 2 for the appropriate species, size, and grade.

The only difference in calculating timber beam deflections arises with long term or permanent loads. Timber will relax under long term loads and acquire a non-elastic set. The standard method of accounting for this long term deflection is to double the calculated deflection for unseasoned members, and to increase the deflections by 50% for seasoned lumber and glue-laminated members.

## D. Bearing Stresses at Loads and Supports

The allowable bearing stress perpendicular to the grain, $F_{c\perp}$, is limited by deformations, not failure. The load duration factors of table 2 do not, therefore, apply to this allowable stress.

When the load is applied at least 3 inches from a member end, and over an area no longer than 6 inches (along the grain), the allowable bearing stress can be multiplied by equation 31.

$$(L_b + .375)/L_b \tag{31}$$

This increase in allowable bearing stress can be applied to the contact area under washers by using a bearing length equal to the washer diameter.

## Example 15

Check the capacity of the glue-laminated crane beam used to carry ore carts. The beam is outdoors and stays wet in service. The only lateral restraint is against rotation at point A. Neglect deflections and assume a normal load duration.

# Timber Design

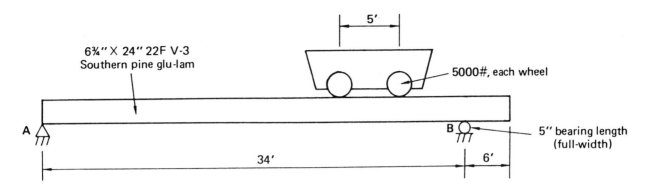

To solve this problem, determine the maximum expected bending and shear stresses. Compare these with the controlling allowable bending stress and the allowable shear stress. Check the bearing stresses at point B.

The maximum bending moment will occur at B with the load at the end of the cantilever, or under the load at some point between A and B. With the cart at the end, the maximum moment is

$$M_{max} = (6) \text{ ft } (5000) \text{ lb} + (1) \text{ ft } (5000) = 35,000 \text{ ft-lb}$$

Using the formula for maximum bending moment with two equal point loads found in most structural handbooks, the maximum moment between the supports is

$$M_{max} = P/2L(L-a/2)^2 = (5000) \text{ lb } [(34) \text{ ft} - (5 \text{ ft}/2)]^2/(2)(34) \text{ ft}$$
$$= 72,960 \text{ ft-lb}$$

Section properties are calculated for glue-laminated members. The section modulus, S, is

$$S = bd^2/6 = (6.75) \text{ in } (24)^2 \text{ in}^2/6 = 648 \text{ in}^3$$

The design bending stresses, therefore, are

$(35,000) \text{ ft-lb } (12) \text{ in/ft/ } (648) \text{ in}^3 = 648 \text{ psi at B}$

$(72,960) \text{ ft-lb } (12) \text{ in/ft/ } (648) \text{ in}^3 = 1,350 \text{ psi between supports}$

The allowable bending stress is the smaller of the stresses limited by lateral buckling and the extreme fiber stress. Lateral buckling is a function of the unbraced length, $L_u$, (here equal to 40 feet,) and the beam configuration. This beam is close to case 4 of figure 40, but to be conservative, use the 1.92 factor. Figure 40 assumes an $L_u/d$ ratio of 17 which may be adjusted. The actual $L_u/d$ ratio is

$$(40) \text{ ft } (12) \text{ in/ft}/(24) \text{ in} = 20.00$$

Thus, the effective length, $L_e$, is

$$(1.92)(40) \text{ ft } (12) \text{ in/ft } [(.85) + (2.55)/(20.00)] = 900 \text{ in}$$

The slenderness factor is

$$C_s = [(900) \text{ in } (24) \text{ in}/(6.75)^2 \text{ in}^2]^{.5} = 21.77$$

This is larger than 10, so the beam is not short. $C_k$ must be calculated to distinguish between long and intermediate beams.

$$C_k = (.811)[(.833)(1{,}600{,}000) \text{ psi}/(.8)(1.0)(2200) \text{ psi}]^{.5}$$
$$= 22.32 > C_s$$

Therefore, the beam is intermediate. The material properties and condition of use factors were found in appendix 2. An adjustment was made for condition of use and load duration when considering lateral buckling. The allowable stress considering lateral buckling is

$$F_b' = F_b (1.0)(0.8)[1 - (1/3)(21.78/22.32)^4] = (0.56) F_b \text{ psi}$$

Now, determine the allowable bending stress considering extreme fiber failure. The load duration factor for normal durations is 1.0. The condition of use factor for wet use is 0.8, from appendix 2. The size factor, $C_F$, is:

$$[(12) \text{ in}/(24) \text{ in}]^{1/9} = .93$$

The allowable bending stress for extreme fiber failure is, therefore,

$$F_b (1.0)(0.8)(.93) = (0.74) F_b \text{ psi}$$

Lateral buckling controls. The allowable bending stress was left in terms of $F_b$ because there will be two table values to evaluate.

The allowable bending stress at B is (.56)(1100) psi = 616 psi, because the bending causes tension at the top of the beam -- normally the compression zone. This is less than the actual 648 psi, so the beam is overstressed in bending at B.

The allowable stress between supports is (.56)(2200) psi = 1232 psi. In this case, the tension is in the tension zone of the beam, resulting in the higher tabulated allowable stress. Again, the actual stress, 1350 psi, is higher than the allowable. With the bending stresses at B also too high, it would be reasonable to specify the 24f-V5 laminae combination, with its balanced construction and higher allowable stresses.

When evaluating the maximum shear force, remember that loads within a beam depth of supports flow directly to the support through compression. The shear force is not maximum, therefore, when the back wheel moves past point B. When the front wheel is at the beam end, the back wheel is still too close to the support to cause shear in the cantilever. The maximum shear, therefore is caused by the cart being 24 inches in from either support. This maximum shear is

(5000) lb [(325.5) in + (385.5) in]/(408) in = 8713 lb

The shear stress is

$f_v$ = 3V/2A = (3)(8714) lb/(2)(6.75) in (24) in = 81 psi

The allowable shear stress, $F_v'$, uses the load duration factor (1.0) of table 2 and the condition of use factor (.875) from appendix 2.

(200) psi (.875)(1.0) = 175 psi,

The allowable shear stress is significantly larger than required.

The maximum support reaction occurs at B when the cart is centered over the support, and this reaction is equal to 10,000 pounds. No load duration factors are applied to compression perpendicular to the grain, but the condition of use factor is .667, found in appendix 2. The allowable bearing stress can be increased because the bearing length is less than 6 inches. The maximum bearing stress is

(10,000) lb /(5) in (6.75) in = 296 psi

The allowable bearing stress, considering equation 31, is

(450) psi (.667) [(5) in + (.375) in]/(5) in = 323 psi

The beam is inadequate in bending and should either be upgraded to a 24f-V5, or increased in size to reduce the bending stresses.

## Example 16

Design a floor girder to carry the loads shown. Use kiln-dried #2 Southern pine. The flooring is nailed to the top of the girder. For occupant satisfaction, deflections are limited to span/360 for live loads. Total deflections should be less than span/240.

The flooring provides full length lateral support, so buckling is not a factor. The bending moments caused by various load combinations will be checked for the critical combination. The critical combination will be used to determine a minimum section modulus. The section will be checked for shear stresses and deflections.

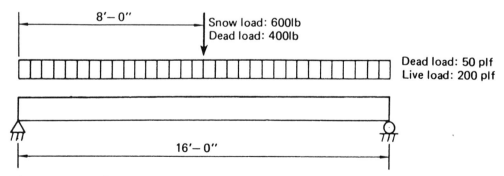

The two logical combinations to check are with and without the snow load. With the snow load included, the maximum moment is

[(50 + 200) plf (16)$^2$ ft$^2$/8 + (400 + 600) lb (16) ft/4](12) in/ft/1.15
= 125,217 in-lb,

# Timber Design

The 1.15 load duration factor of table 2 is applied because the load combination includes the snow load. The net moment without the snow load is

$$[(50 + 200) \text{ plf } (16)^2 \text{ ft}^2/8 + (400) \text{ lb } (16) \text{ ft}/4](12) \text{ in/ft}$$
$$= 115{,}200 \text{ in-lb}$$

Even though the snow load combination controls, the stresses are nearly as critical with the snow load left off because the lower load has a longer duration.

Appendix 2 gives a single use $F_b$ of 1300 psi for #2 KD Southern pine. Thus, the required section modulus is

$$S_{required} = (125{,}217) \text{ in-lb}/(1300) \text{ psi} = 96.32 \text{ in}^3$$

Note that the load duration factor is not included in this calculation -- it is already in the design bending moment.

Using the section properties of appendix 5, the required section modulus can be obtained in various ways.

$$(96.32) \text{ in}^3/(21.391) \text{ in}^3 = 4.50; \text{ 5-2x10's, or,}$$
$$(96.32) \text{ in}^3/(31.641) \text{ in}^3 = 3.04; \text{ 3-2x12's}$$

However, with three interconnected side-by-side members, the repetitive use allowable bending stress is justified. Appendix 2 gives 1500 psi for these conditions. Therefore, the required section modulus is only

$$S_{required} = (125{,}217) \text{ in-lb}/(1500) \text{ psi} = 83.48 \text{ in}^3$$

This section modulus is available in various forms.

$$(83.48) \text{ in}^3/(21.391) \text{ in}^3 = 3.90; \text{ 4-2x10's, or,}$$
$$(83.48) \text{ in}^3/(31.641) \text{ in}^3 = 2.64; \text{ 3-2x12's}$$

The 3-2x12's require less area (lumber). If the shear stress is not too high, and if the cost is not higher for the larger size, pick the 2x12's. They will also provide greater stiffness.

To check the shear stress, determine the critical shear force. Neglect the distributed load within a beam depth of the supports. First, calculate the shear force with the snow load.

$$V_{max} = (600 + 400) \text{ lb}/2 + (50 + 200) \text{ plf } [(16)/2 - (11.25)/12] \text{ ft}$$
$$= 2266 \text{ lb}/(1.15)$$
$$= 1970 \text{ lb}$$

Now, calculate the shear without the snow load

$$V_{max} = (400) \text{ lb}/2 + (50 + 200) \text{ plf } [(16)/2 - (11.25)/12] \text{ ft}$$
$$= 1966 \text{ lb}$$

The maximum shear stress is

$$f_v = (1.5)(1970) \text{ lb}/(3)(16.875) \text{ in}^2 = 58.4 \text{ psi}$$

The allowable stress is 95 psi, so the 3-2x12's are adequate. The allowable shear stress could have been increased by 50% with the carefully checked maximum shear forces, but this was not required.

Now check the deflections. Use tabled formulae for maximum deflections in simple spans. Use half the modulus of elasticity for the dead load component to account for long term creep effects.

$$D_{dead} = \frac{\frac{(5)(50/12)[(16)(12)]^4}{384} + \frac{(400)[(16)(12)]^3}{48}}{(1,600,000/2)(3)(177.979)}$$

$$= 0.31"$$

$$D_{live} = \frac{\frac{(5)(200/12)\text{lb/in}[(16)(12)]^4}{348} + \frac{600 \text{ lb }[(16)(12)]^3}{48}}{1,600,000 \text{ psi }(3)(177.979)\text{in}^4}$$

$$= 0.45" = \text{span}/[(16)(12)/(.45) = \text{span}/457 < 1/360,$$

The total deflection is (.31) + (.45) = .76", equal to span/250, slightly less than the maximum allowable deflection of span/240. So, the 3-2x12, #2 Southern pine girder is adequate.

## 10. PLYWOOD

### A. Introduction

Plywood is a product with many varied structural applications. Plywood is valuable because of its size and properties. A piece of solid sawn wood 1/2"x48"x96" would be very expensive and absolutely useless structurally, but the same size piece of plywood is much cheaper and, in some ways, stronger than the wood of which it is made.

Virtually any softwood species can be used in plywood. Plywood is made by peeling logs into veneer and laminating layers of this veneer with glue. The strength results from cutting up and spreading out the knots and other defects, and cross-banding the laminae. Cross-banding means that alternate laminae are perpendicular to each other. The result is that plywood has two strong directions instead of one.

#### Plywood Grades and Types

Plywood is available in many forms and grades. The variables in plywood composition are veneer grade and species, veneer configuration, and glue type. Structural plywood veneers are graded from A to D. Plywood is designated by the veneer grades that appear in the outer or face plies. C-D plywood, for example, has a D grade veneer on the back face and a C grade face with smaller knot holes.

With all the potential variations in plywood composition, it is only through the standards established by the American Plywood Association[3] that the end users can know what to expect from a given sheet. The standards are written as fabrication limits for equivalent plywood designations.

---

[3] American Plywood Association, 1119 A Street, Tacoma, Washington 98401

A given plywood thickness can have various combinations of the number of plies and their individual thicknesses. In marine grade plywood, the veneers are oriented at other angles than just the perpendicular, giving the product an even more uniform performance in all directions. The two major glue types are interior and exterior, depending on the glue's resistance to wet-dry cycles.

Plywood Structural Applications

The most common plywood applications are flooring, roofing, and siding. The plywood spans the space between joists, rafters, or studs, and distributes loads to those members. Engineering calculations can be used to investigate the plywood stresses under these conditions, but it is much simpler to follow the allowable span recommendations found in the certification stamp for most common decking plywoods. Figure 43 is an example plywood stamp. The 24/0 means that the maximum roof rafter spacing is 24 inches, and that the plywood is not recommended for floor decking with any spacing.

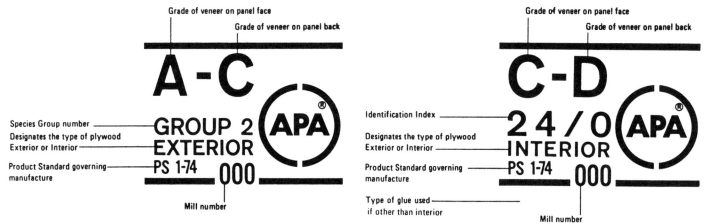

**Figure 43**
**APA Grade-Trademark Stamp**
By permission of APA

Plywood sheathing on walls, roofs, and floors can resist more than just loads normal to the surface. If the plywood is sized and connected adequately, the walls, roofs, and floors can act as shear walls or diaphragms in resisting lateral loads on a building. Sample problem 17 deals with diaphragm design.

Plywood is very strong and rigid against forces that tend to distort it out-of-square. This strength and its shape makes plywood a logical choice for use as the web of a built-up beam. Box and I-beams can be fabricated with lumber flanges. Sample problem 18 involves sorting through all the variables to design a plywood-lumber beam.

It is possible to make an efficient double use of plywood sheathing in stressed-skin or sandwich panels. In these structural elements, the plywood skin acts as the flanges of a shallow, wide beam. Shear resistance is provided by the lumber webs or sandwich core material. Supplements to the Plywood Design Specifications, published by the APA, provide the information needed to design and analyze these relatively sophisticated elements.

## B. Plywood Section Properties

### Direction of Face Grain

In plywood, alternating plies are perpendicular to each other. Since wood has markedly different properties across and along the grain, it is necessary to know which direction the plies are oriented. The standard orientation reference is the grain direction of the visible face plies. The grain in the face plies of a 4'x8' plywood panel almost always runs in the 8' direction.

The effective section properties of table 20 depend on the orientation of the stresses relative to this face grain direction. Plywood used as sheathing is strongest against normal loads if the face grain is across the supporting members. Plywood loaded in this strong direction is an example of the "Stress Applied Parallel to Face Grain" category used in table 20.

### Thickness for All Properties Except Shear

For all calculations other than those involving shear, the nominal thickness found in column 1 of table 20 is used.

### Thickness for Shear

For calculating shear stresses, the effective thicknesses of column 3 in table 20 are used. For structural plywood grades, this effective thickness can be larger than the actual, nominal thickness.

## Table 20
## Effective Plywood Section Properties

**Face Plies of Different Species Group from Inner Plies**

| Nominal Thickness (in.) | Approximate Weight (psf) | $t_s$ Effective Thickness For Shear (in.) | Stress Applied Parallel to Face Grain | | | | Stress Applied Perpendicular to Face Grain | | | |
|---|---|---|---|---|---|---|---|---|---|---|
| | | | A Area (in.$^2$/ft) | I Moment of Inertia (in.$^4$/ft) | KS Effective Section Modulus (in.$^3$/ft) | Ib/Q Rolling Shear Constant (in.$^2$/ft) | A Area (in.$^2$/ft) | I Moment of Inertia (in.$^4$/ft) | KS Effective Section Modulus (in.$^3$/ft) | Ib/Q Rolling Shear Constant (in.$^2$/ft) |
| **UNSANDED PANELS** | | | | | | | | | | |
| 5/16-U | 1.0 | 0.268 | 1.491 | 0.022 | 0.112 | 2.569 | 0.660 | 0.001 | 0.023 | 4.497 |
| 3/8 -U | 1.1 | 0.278 | 1.866 | 0.039 | 0.152 | 3.110 | 0.799 | 0.002 | 0.033 | 5.444 |
| 15/32 & 1/2 -U | 1.5 | 0.298 | 2.292 | 0.067 | 0.213 | 3.921 | 1.007 | 0.004 | 0.056 | 2.450 |
| 19/32 & 5/8 -U | 1.8 | 0.319 | 2.330 | 0.121 | 0.379 | 5.004 | 1.285 | 0.010 | 0.091 | 3.106 |
| 23/32 & 3/4 -U | 2.2 | 0.445 | 3.247 | 0.234 | 0.496 | 6.455 | 1.563 | 0.036 | 0.232 | 3.613 |
| 7/8 -U | 2.6 | 0.607 | 3.509 | 0.340 | 0.678 | 7.175 | 1.950 | 0.112 | 0.397 | 4.791 |
| 1 -U | 3.0 | 0.842 | 3.916 | 0.493 | 0.859 | 9.244 | 3.145 | 0.210 | 0.660 | 6.533 |
| 1-1/8 -U | 3.3 | 0.859 | 4.725 | 0.676 | 1.047 | 9.960 | 3.079 | 0.288 | 0.768 | 7.931 |
| **SANDED PANELS** | | | | | | | | | | |
| 1/4 -S | 0.8 | 0.267 | 0.996 | 0.008 | 0.059 | 2.010 | 0.348 | 0.001 | 0.009 | 2.019 |
| 11/32-S | 1.0 | 0.284 | 0.996 | 0.019 | 0.093 | 2.765 | 0.417 | 0.001 | 0.016 | 2.589 |
| 3/8 -S | 1.1 | 0.288 | 1.307 | 0.027 | 0.125 | 3.088 | 0.626 | 0.002 | 0.023 | 3.510 |
| 15/32-S | 1.4 | 0.421 | 1.947 | 0.066 | 0.214 | 4.113 | 1.204 | 0.006 | 0.067 | 2.434 |
| 1/2 -S | 1.5 | 0.425 | 1.947 | 0.077 | 0.236 | 4.466 | 1.240 | 0.009 | 0.087 | 2.752 |
| 19/32-S | 1.7 | 0.546 | 2.423 | 0.115 | 0.315 | 5.471 | 1.389 | 0.021 | 0.137 | 2.861 |
| 5/8 -S | 1.8 | 0.550 | 2.475 | 0.129 | 0.339 | 5.824 | 1.528 | 0.027 | 0.164 | 3.119 |
| 23/32-S | 2.1 | 0.563 | 2.822 | 0.179 | 0.389 | 6.581 | 1.737 | 0.050 | 0.231 | 3.818 |
| 3/4 -S | 2.2 | 0.568 | 2.884 | 0.197 | 0.412 | 6.762 | 2.081 | 0.063 | 0.285 | 4.079 |
| 7/8 -S | 2.6 | 0.586 | 2.942 | 0.278 | 0.515 | 8.050 | 2.651 | 0.104 | 0.394 | 5.078 |
| 1 -S | 3.0 | 0.817 | 3.721 | 0.423 | 0.664 | 8.882 | 3.163 | 0.185 | 0.591 | 7.031 |
| 1-1/8 -S | 3.3 | 0.836 | 3.854 | 0.548 | 0.820 | 9.883 | 3.180 | 0.271 | 0.744 | 8.428 |
| **TOUCH-SANDED PANELS** | | | | | | | | | | |
| 1/2 -T | 1.5 | 0.342 | 2.698 | 0.083 | 0.271 | 4.252 | 1.159 | 0.006 | 0.061 | 2.746 |
| 19/32 & 5/8 -T | 1.8 | 0.408 | 2.354 | 0.123 | 0.327 | 5.346 | 1.555 | 0.016 | 0.135 | 3.220 |
| 23/32 & 3/4 -T | 2.2 | 0.439 | 2.715 | 0.193 | 0.398 | 6.589 | 1.622 | 0.032 | 0.219 | 3.635 |
| 1-1/8 -T | 3.3 | 0.839 | 4.548 | 0.633 | 0.977 | 11.258 | 4.067 | 0.272 | 0.743 | 8.535 |

By permission of APA

## Structural I and Marine

| Nominal Thickness (in.) | Approximate Weight (psf) | $t_s$ Effective Thickness For Shear (in.) | Stress Applied Parallel to Face Grain ||||  Stress Applied Perpendicular to Face Grain ||||
|---|---|---|---|---|---|---|---|---|---|---|
| | | | $A$ Area (in.$^2$/ft) | $I$ Moment of Inertia (in.$^4$/ft) | $KS$ Effective Section Modulus (in.$^3$/ft) | $Ib/Q$ Rolling Shear Constant (in.$^2$/ft) | $A$ Area (in.$^2$/ft) | $I$ Moment of Inertia (in.$^4$/ft) | $KS$ Effective Section Modulus (in.$^3$/ft) | $Ib/Q$ Rolling Shear Constant (in.$^2$/ft) |
| UNSANDED PANELS |||||||||||
| 5/16-U | 1.0 | 0.356 | 1.619 | 0.022 | 0.126 | 2.567 | 1.188 | 0.002 | 0.029 | 6.037 |
| 3/8 -U | 1.1 | 0.371 | 2.226 | 0.041 | 0.195 | 3.107 | 1.438 | 0.003 | 0.043 | 7.307 |
| 15/32 & 1/2 -U | 1.5 | 0.535 | 2.719 | 0.074 | 0.279 | 4.157 | 2.175 | 0.012 | 0.116 | 2.408 |
| 19/32 & 5/8 -U | 1.8 | 0.707 | 3.464 | 0.154 | 0.437 | 5.685 | 2.742 | 0.045 | 0.240 | 3.072 |
| 23/32 & 3/4 -U | 2.2 | 0.739 | 4.219 | 0.236 | 0.549 | 6.148 | 2.813 | 0.064 | 0.299 | 3.540 |
| 7/8 -U | 2.6 | 0.776 | 4.388 | 0.346 | 0.690 | 6.948 | 3.510 | 0.131 | 0.457 | 4.722 |
| 1 -U | 3.0 | 1.088 | 5.200 | 0.529 | 0.922 | 8.512 | 5.661 | 0.270 | 0.781 | 6.435 |
| 1-1/8 -U | 3.3 | 1.118 | 6.654 | 0.751 | 1.164 | 9.061 | 5.542 | 0.408 | 0.999 | 7.833 |
| SANDED PANELS |||||||||||
| 1/4 -S | 0.8 | 0.342 | 1.280 | 0.012 | 0.083 | 2.009 | 0.626 | 0.001 | 0.013 | 2.723 |
| 11/32-S | 1.0 | 0.365 | 1.280 | 0.026 | 0.133 | 2.764 | 0.751 | 0.001 | 0.023 | 3.397 |
| 3/8 -S | 1.1 | 0.373 | 1.680 | 0.038 | 0.177 | 3.086 | 1.126 | 0.002 | 0.033 | 4.927 |
| 15/32-S | 1.4 | 0.537 | 1.947 | 0.067 | 0.246 | 4.107 | 2.168 | 0.009 | 0.093 | 2.405 |
| 1/2 -S | 1.5 | 0.545 | 1.947 | 0.078 | 0.271 | 4.457 | 2.232 | 0.014 | 0.123 | 2.725 |
| 19/32-S | 1.7 | 0.709 | 3.018 | 0.116 | 0.338 | 5.566 | 2.501 | 0.034 | 0.199 | 2.811 |
| 5/8 -S | 1.8 | 0.717 | 3.112 | 0.131 | 0.361 | 5.934 | 2.751 | 0.045 | 0.238 | 3.073 |
| 23/32-S | 2.1 | 0.741 | 3.735 | 0.183 | 0.439 | 6.109 | 3.126 | 0.085 | 0.338 | 3.780 |
| 3/4 -S | 2.2 | 0.748 | 3.848 | 0.202 | 0.464 | 6.189 | 3.745 | 0.108 | 0.418 | 4.047 |
| 7/8 -S | 2.6 | 0.778 | 3.952 | 0.288 | 0.569 | 7.539 | 4.772 | 0.179 | 0.579 | 5.046 |
| 1 -S | 3.0 | 1.091 | 5.215 | 0.479 | 0.827 | 7.978 | 5.693 | 0.321 | 0.870 | 6.981 |
| 1-1/8 -S | 3.3 | 1.121 | 5.593 | 0.623 | 0.955 | 8.841 | 5.724 | 0.474 | 1.098 | 8.377 |
| TOUCH-SANDED PANELS |||||||||||
| 1/2 -T | 1.5 | 0.543 | 2.698 | 0.084 | 0.282 | 4.511 | 2.486 | 0.020 | 0.162 | 2.720 |
| 19/32 & 5/8 -T | 1.8 | 0.707 | 3.127 | 0.124 | 0.349 | 5.500 | 2.799 | 0.050 | 0.259 | 3.183 |
| 23/32 & 3/4 -T | 2.2 | 0.739 | 4.059 | 0.201 | 0.469 | 6.592 | 3.625 | 0.078 | 0.350 | 3.596 |

By permission of APA

### Cross Sectional Area

The effective areas of columns 4 and 8 in table 20 reflect the differences in grain orientation. Those plies whose fibers are perpendicular to the direction of stress application are neglected as having essentially no stiffness or strength. Note that the effective areas are much larger when the stress is applied parallel to the face.

The effective area is given in units of $in^2/ft$. This per foot basis is also common to the rest of the section properties. The "foot" is a twelve inch wide strip of the specific piece of plywood, measured across the direction of stress application. A 4x8 panel with the stress parallel to the face grain would, therefore, have an effective area resisting that stress of four times the value found in column 4 of table 20.

### Moment of Inertia

Only those plies whose fibers are parallel to the stress direction are included in the values of columns 5 and 10 of table 20. The per foot unit means the same as it does for effective area.

### Section Modulus

The effective section modulus, KS, includes an empirical factor K. This value should always be used for bending stress calculations when the plywood is installed as sheathing and loaded normal to the surface. Specifically, the apparently valid I/c value for an effective section modulus should not be used. Figure 44 illustrates the difference between loading in the plane of the plywood and loading normal to that plane.

When plywood is used as the web of a plywood-lumber beam, bending stresses in the plywood are more like axial forces. A different section modulus will be used in those calculations.

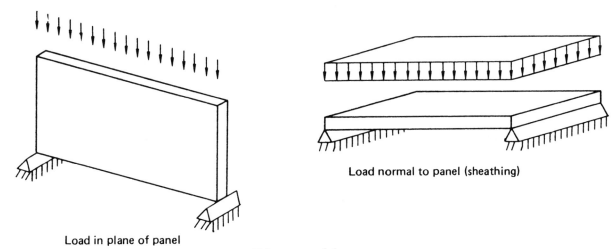

**Figure 44
Plywood in Bending**

### Rolling Shear Constant

When plywood is loaded in shear through the thickness (figure 45), it is tremendously strong because the cross plies are being sheared across the grain. When the shear stresses lie in the plane of the plies, however, plywood is not nearly as strong. The rolling shear that developes (figure 45) tends to roll the fibers of the cross plies over each other, a tendency against which wood has little resistance.

Rolling shear stresses arise in plywood used as sheathing, and at the connection between plywood webs and lumber flanges in plywood-lumber beams.

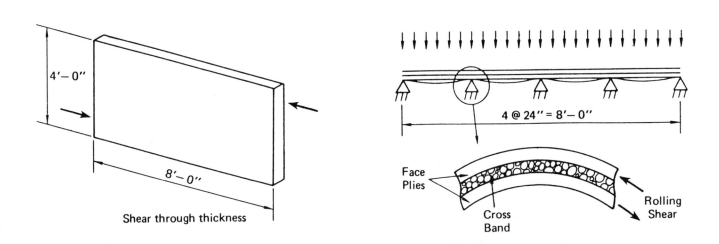

**Figure 45
Shear Stress Orientations**

## C. Allowable Stresses -- Plywood

There are two basic factors which determine the allowable stresses in plywood: the species used in the laminae, and how those laminae were fabricated. The allowable stresses are further subject to modification for load duration and conditions of use.

### Grade Stress Levels

The first distinction drawn in the allowable plywood stress tables is grade of the laminae used in the plywood. The grades are classified in three grade stress levels: S-1, S-2, and S-3. Table 21 is a listing of the most common structural plywood types. Once the plywood type is specified, table 21 will give the grade stress level.

### Plywood Species Groups

The next distinction required to determine the allowable stress is the species used to make the laminae. For certain plywoods, table 21 contains this information also. If table 21 does not specify the species group used in a plywood, there are two other ways to determine the information.

If the species is known, table 22 provides the classification information.

Very often, however, the plywood manufacturer will not stamp the plywood with the species. The stamp could include an allowable roof/floor span rating which can be used to indirectly determine the species. Figure 46 gives the species group as a function of the plywood thickness and "span rating". For example, a 5/8" piece of plywood with group 3 species plies, and a 1/2" sheet of group 1 species plywood will both handle 32" rafter spacings and 16" joist spacings.

# Timber Design

## Table 21
## Plywood Specifications -- Abbreviated

### GUIDE TO USE OF ALLOWABLE STRESS AND SECTION PROPERTIES TABLES

| | Plywood Grade | Description and Use | Typical Trademarks | Veneer Grade | | | Common Thicknesses | Grade Stress Level (Table 3) | Species Group | Section Property Table |
|---|---|---|---|---|---|---|---|---|---|---|
| | | | | Face | Back | Inner | | | | |
| **INTERIOR OR PROTECTED APPLICATIONS** | APA RATED SHEATHING EXP 1 or 2[3] | Unsanded sheathing grade for wall, roof, sub-flooring, and industrial applications such as pallets and for engineering design, with proper stresses. Manufactured with intermediate and exterior glue (1). For permanent exposure to weather or moisture only Exterior type plywood is suitable. | APA RATED SHEATHING 32/16 15/32 INCH SIZED FOR SPACING EXPOSURE 1 000 PS 1-83 C-D NER-108 | C | D | D | 5/16, 3/8, 15/32, 1/2, 19/32, 5/8, 23/32, 3/4 | S-3 (1) | See "Key to Span Rating" | Table 1 (unsanded) |
| | APA STRUCTURAL I RATED SHEATHING EXP 1 or APA STRUCTURAL II[2] RATED SHEATHING EXP 1 | Plywood grades to use where strength properties are of maximum importance, such as plywood-lumber components. Made with exterior glue only. STRUCTURAL I is made from all Group 1 woods. STRUCTURAL II allows Group 3 woods. | APA RATED SHEATHING STRUCTURAL I 24/0 3/8 INCH SIZED FOR SPACING EXPOSURE 1 000 PS 1-83 C-D NER-108 | C | D | D | 5/16, 3/8, 15/32, 1/2, 19/32, 5/8, 23/32, 3/4 | S-2 | Structural I use Group 1 | Table 2 (unsanded) |
| | | | | | | | | | Structural II See "Key to Span Rating" | Table 1 (unsanded) |
| | APA RATED STURD-I-FLOOR EXP 1 or 2[3] | For combination subfloor-underlayment. Provides smooth surface for application of carpet. Possesses high concentrated and impact load resistance during construction and occupancy. Manufactured with intermediate and exterior glue. Touch-sanded (4). Available with tongue and groove.(5) | APA RATED STURD-I-FLOOR 20 OC 19/32 INCH SIZED FOR SPACING T&G NET WIDTH 47-1/2 EXPOSURE 1 000 UNDERLAYMENT PS 1-83 NER-108 | C plugged | D | C & D | 19/32, 5/8, 23/32, 3/4, 1-1/8 (2-4-1) | S-3 (1) | See "Key to Span Rating" | Table 1 (touch-sanded) |
| | APA UNDERLAYMENT EXP 1, 2 or INT | For underlayment under carpet. Available with exterior glue. Touch-sanded. Available with tongue and groove.(5) | APA UNDERLAYMENT GROUP 1 EXPOSURE 1 000 PS 1-83 | C plugged | D | C & D | 1/2, 19/32, 5/8, 23/32, 3/4 | S-3 (1) | As Specified | Table 1 (touch-sanded) |
| | APA C-D PLUGGED EXP 1, 2 or INT | For built-ins, wall and ceiling tile backing, NOT for underlayment. Available with exterior glue. Touch-sanded.(5) | APA C-D PLUGGED GROUP 2 EXPOSURE 1 000 PS 1-83 | C plugged | D | D | 1/2, 19/32, 5/8, 23/32, 3/4 | S-3 (1) | As Specified | Table 1 (touch-sanded) |
| | APA APPEARANCE GRADES EXP 1, 2 or INT | Generally applied where a high quality surface is required. Includes APA N-N, N-A, N-B, N-D, A-A, A-B, A-D, B-B, and B-D INT grades.(5) | APA A-D GROUP 1 EXPOSURE 1 000 PS 1-83 | B or better | D or better | C & D | 1/4, 11/32, 3/8, 15/32, 1/2, 19/32, 5/8, 23/32, 3/4 | S-3 (1) | As Specified | Table 1 (sanded) |

(1) When exterior glue is specified, i.e. Exposure 1, stress level 2 (S-2) should be used.
(2) Check local suppliers for availability of STRUCTURAL II and PLYFORM Class II grades.
(3) Properties and stresses apply only to APA RATED STURD-I-FLOOR and APA RATED SHEATHING manufactured entirely with veneers.
(4) APA RATED STURD-I-FLOOR 2-4-1 may be produced unsanded.
(5) May be modified to STRUCTURAL I. For such designation use Group 1 stresses and Table 2 section properties.

By permission of APA

## GUIDE TO USE OF ALLOWABLE STRESS AND SECTION PROPERTIES TABLES (Continued)

| | Plywood Grade | Description and Use | Typical Trademarks | Veneer Grade Face | Veneer Grade Back | Veneer Grade Inner | Common Thicknesses | Grade Stress Level (Table 3) | Species Group | Section Property Table |
|---|---|---|---|---|---|---|---|---|---|---|
| **EXTERIOR APPLICATIONS** | APA RATED SHEATHING EXT[3] | Unsanded sheathing grade with waterproof glue bond for wall, roof, subfloor and industrial applications such as pallet bins. | APA RATED SHEATHING 48/24 3/4 INCH SIZED FOR SPACING EXTERIOR 000 PS 1-74 C-C NRB-108 | C | C | C | 5/16, 3/8, 1/2, 5/8, 3/4 | S-1 | See "Key to Span Rating" | Table 1 (unsanded) |
| | APA STRUCTURAL I RATED SHEATHING EXT or APA STRUCTURAL II[2] RATED SHEATHING EXT | "Structural" is a modifier for this unsanded sheathing grade. For engineered applications in construction and industry where full exterior-type panels are required. STRUCTURAL I is made from Group 1 woods only. | APA RATED SHEATHING STRUCTURAL I 24/0 3/8 INCH SIZED FOR SPACING EXTERIOR 000 PS 1-74 C-C NRB-108 | C | C | C | 5/16, 3/8, 1/2, 5/8, 3/4 | S-1 | Structural I use Group 1 | Table 2 (unsanded) |
| | | | | | | | | | Structural II See "Key to Span Rating" | Table 1 (unsanded) |
| | APA RATED STURD-I-FLOOR EXT[3] | For combination subfloor-underlayment for resilient floor coverings where severe moisture conditions may be present, as in balcony decks. Possesses high concentrated and impact load resistance during construction and occupancy. Touch-sanded.[4] Available with tongue and groove.[5] | APA RATED STURD-I-FLOOR 20 OC 19/32 INCH SIZED FOR SPACING EXTERIOR 000 NRB-108 FHA-UM-65 | C plugged | C | C | 19/32, 5/8, 23/32, 3/4, | S-2 | See "Key to Span Rating" | Table 1 (touch-sanded) |
| | APA UNDER-LAYMENT EXT and APA C-C PLUGGED EXT | Underlayment for floor under resilient floor coverings where severe moisture conditions may exist. Also for controlled atmosphere rooms and many industrial applications. Touch-sanded. Available with tongue and groove.[5] | APA C-C PLUGGED GROUP 2 EXTERIOR 000 PS 1-74 | C plugged | C | C | 1/2, 19/32, 5/8, 23/32, 3/4 | S-2 | As Specified | Table 1 (touch-sanded) |
| | APA B-B PLYFORM CLASS I or II[2] | Concrete-form grade with high reuse factor. Sanded both sides, mill-oiled unless otherwise specified. Available in HDO. For refined design information on this special-use panel see APA publication "Plywood for Concrete Forming" (form V345). Design using values from this specification will result in a conservative design.[5] | APA PLYFORM B-B CLASS I EXTERIOR 000 PS 1-74 | B | B | C | 5/8, 3/4 | S-2 | Class I use Group 1; Class II use Group 3 | Table 1 (sanded) |
| | APA MARINE EXT | Superior Exterior-type plywood made only with Douglas Fir or Western Larch. Special solid-core construction. Available with MDO or HDO face. Ideal for boat hull construction. | MARINE A-A EXT APA PS 1-74 000 | A or B | A or B | B | 1/4, 3/8, 1/2, 5/8, 3/4 | A face & back use S-1; B face or back use S-2 | Group 1 | Table 2 (sanded) |
| | APA APPEARANCE GRADES EXT | Generally applied where a high quality surface is required. Includes APA A-A, A-B, A-C, B-B, B-C, HDO and MDO EXT.[5] | APA A-C GROUP 1 EXTERIOR 000 PS 1-74 | B or better | C or better | C | 1/4, 3/8, 1/2, 5/8, 3/4 | A or C face and back use S-1; B face or back use S-2 | As Specified | Table 1 (sanded) |

(1) When exterior glue is specified, i.e. "Interior with exterior glue" or "Exposure 1," stress level 2 (S-2) should be used.
(2) Check local suppliers for availability of STRUCTURAL II and PLYFORM Class II grades.
(3) Properties and stresses apply only to APA RATED STURD-I-FLOOR and APA RATED SHEATHING manufactured entirely with veneers.
(4) APA RATED STURD-I-FLOOR 2-4-1 may be produced unsanded.
(5) May be modified to STRUCTURAL I. For such designation use Group 1 stresses and Table 2 section properties.

By permission of APA

# TIMBER DESIGN

## Table 22
## Plywood Species Classification

| Group 1 | Group 2 | | Group 3 | Group 4 | Group 5[a] |
|---------|---------|---|---------|---------|-----------|
| Apitong[b][c]<br>Beech, American<br>Birch<br>  Sweet<br>  Yellow<br>Douglas Fir 1[d]<br>Kapur[b]<br>Keruing[b][c]<br>Larch, Western<br>Maple, Sugar<br>Pine<br>  Caribbean<br>  Ocote<br>Pine, Southern<br>  Loblolly<br>  Longleaf<br>  Shortleaf<br>  Slash<br>Tanoak | Cedar, Port Orford<br>Cypress<br>Douglas Fir 2[d]<br>Fir<br>  Balsam<br>  California Red<br>  Grand<br>  Noble<br>  Pacific Silver<br>  White<br>Hemlock, Western<br>Lauan<br>  Almon<br>  Bagtikan<br>  Mayapis<br>  Red Lauan<br>  Tangile<br>  White Lauan | Maple, Black<br>Mengkulang[b]<br>Meranti, Red[b][e]<br>Mersawa[b]<br>Pine<br>  Pond<br>  Red<br>  Virginia<br>  Western White<br>Spruce<br>  Black<br>  Red<br>  Sitka<br>Sweetgum<br>Tamarack<br>Yellow-poplar | Alder, Red<br>Birch, Paper<br>Cedar, Alaska<br>Fir, Subalpine<br>Hemlock, Eastern<br>Maple, Bigleaf<br>Pine<br>  Jack<br>  Lodgepole<br>  Ponderosa<br>  Spruce<br>Redwood<br>Spruce<br>  Engelmann<br>  White | Aspen<br>  Bigtooth<br>  Quaking<br>Cativo<br>Cedar<br>  Incense<br>  Western Red<br>Cottonwood<br>  Eastern<br>  Black (Western<br>    Poplar)<br>Pine<br>  Eastern White<br>  Sugar | Basswood<br>Poplar, Balsam |

(a) Design stresses for Group 5 not assigned.
(b) Each of these names represents a trade group of woods consisting of a number of closely related species.
(c) Species from the genus Dipterocarpus are marketed collectively: Apitong if originating in the Philippines; Keruing if originating in Malaysia or Indonesia.
(d) Douglas fir from trees grown in the states of Washington, Oregon, California, Idaho, Montana, Wyoming, and the Canadian Provinces of Alberta and British Columbia shall be classed as Douglas fir No. 1. Douglas fir from trees grown in the states of Nevada, Utah, Colorado, Arizona and New Mexico shall be classed as Douglas fir No. 2.
(e) Red Meranti shall be limited to species having a specific gravity of 0.41 or more based on green volume and oven dry weight.

By permission of APA

## Key to Span Rating and Species Group

For panels with "Span Rating" as across top, and thickness as at left, use stress for species group given in table.

| Thickness (in.) | Span Rating (APA RATED SHEATHING grades) | | | | | | | |
|---|---|---|---|---|---|---|---|---|
| | 12/0 | 16/0 | 20/0 | 24/0 | 32/16 | 40/20 | 48/24 | |
| | | | | | Span Rating (STURD-I-FLOOR grades) | | | |
| | | | | | 16 o.c. | 20 o.c. | 24 o.c. | 48 o.c. |
| 5/16 | 4 | 3 | 1 | | | | | |
| 3/8 | | | 4[3] | 1 | | | | |
| 15/32 & 1/2 | | | | 4[3] | 1[1] | | | |
| 19/32 & 5/8 | | | | | 4[3] | 1 | | |
| 23/32 & 3/4 | | | | | | 4[3] | 1 | |
| 7/8 | | | | | | | 3[2] | |
| 1-1/8 | | | | | | | | 1 |

(1) Thicknesses not applicable to APA RATED STURD-I-FLOOR.
(2) For APA RATED STURD-I-FLOOR 24 oc, use Group 4 stresses.
(3) For STRUCTURAL II, use Group 3 stresses.

### Figure 46
### Species Group -- Span Rating Relationship

By permission of APA

## Table 23
### Plywood Allowable Stresses

**Allowable Stresses for Plywood (psi)** conforming to U.S. Product Standard PS 1-74/ANSI A199.1 for Construction and Industrial Plywood. Stresses are based on normal duration of load, and on common structural applications where panels are 24" or greater in width. For other use conditions, see Section 3.3 for modifications.

| Type of Stress | Species Group of Face Ply | Grade Stress Level[1] | | | | |
|---|---|---|---|---|---|---|
| | | S-1 | | S-2 | | S-3 |
| | | Wet | Dry | Wet | Dry | Dry Only |
| **EXTREME FIBER STRESS IN BENDING ($F_b$)** $F_b$ | 1 | 1430 | 2000 | 1190 | 1650 | 1650 |
| **TENSION IN PLANE OF PLIES ($F_t$)** & $F_t$ | 2, 3 | 980 | 1400 | 820 | 1200 | 1200 |
| Face Grain Parallel or Perpendicular to Span | 4 | 940 | 1330 | 780 | 1110 | 1110 |
| (At 45° to Face Grain Use 1/6 $F_t$) | | | | | | |
| **COMPRESSION IN PLANE OF PLIES** | 1 | 970 | 1640 | 900 | 1540 | 1540 |
| | 2 | 730 | 1200 | 680 | 1100 | 1100 |
| Parallel or Perpendicular to Face Grain $F_c$ | 3 | 610 | 1060 | 580 | 990 | 990 |
| (At 45° to Face Grain Use 1/3 $F_c$) | 4 | 610 | 1000 | 580 | 950 | 950 |
| **SHEAR THROUGH THE THICKNESS**[3] | 1 | 155 | 190 | 155 | 190 | 160 |
| Parallel or Perpendicular to Face Grain $F_v$ | 2, 3 | 120 | 140 | 120 | 140 | 120 |
| (At 45° to Face Grain Use 2 $F_v$) | 4 | 110 | 130 | 110 | 130 | 115 |
| **ROLLING SHEAR (IN THE PLANE OF PLIES)** | MARINE & STRUCTURAL I | 63 | 75 | 63 | 75 | — |
| Parallel or Perpendicular to Face Grain $F_s$ | ALL OTHER[2] | 44 | 53 | 44 | 53 | 48 |
| (At 45° to Face Grain Use 1-1/3 $F_s$) | | | | | | |
| **MODULUS OF RIGIDITY** | 1 | 70,000 | 90,000 | 70,000 | 90,000 | 82,000 |
| Shear in Plane Perpendicular to Plies $G$ | 2 | 60,000 | 75,000 | 60,000 | 75,000 | 68,000 |
| | 3 | 50,000 | 60,000 | 50,000 | 60,000 | 55,000 |
| | 4 | 45,000 | 50,000 | 45,000 | 50,000 | 45,000 |
| **BEARING (ON FACE)** | 1 | 210 | 340 | 210 | 340 | 340 |
| Perpendicular to Plane of Plies $F_{c\perp}$ | 2, 3 | 135 | 210 | 135 | 210 | 210 |
| | 4 | 105 | 160 | 105 | 160 | 160 |
| **MODULUS OF ELASTICITY IN BENDING IN PLANE OF PLIES** $E$ | 1 | 1,500,000 | 1,800,000 | 1,500,000 | 1,800,000 | 1,800,000 |
| | 2 | 1,300,000 | 1,500,000 | 1,300,000 | 1,500,000 | 1,500,000 |
| Face Grain Parallel or Perpendicular to Span | 3 | 1,100,000 | 1,200,000 | 1,100,000 | 1,200,000 | 1,200,000 |
| | 4 | 900,000 | 1,000,000 | 900,000 | 1,000,000 | 1,000,000 |

(1) See pages 14 and 15 for Guide.
To qualify for stress level S-1, gluelines must be exterior and only veneer grades N, A, and C are allowed in either face or back. For stress level S-2, gluelines must be exterior and veneer grade B, C-Plugged and D are allowed on the face or back. Stress level S-3 includes all panels with interior or intermediate gluelines.
(2) Reduce stresses 25% for 3-layer (4-ply) panels over 5/8" thick. Such layups are possible under PS 1-74 for APA RATED SHEATHING, APA RATED STURD-I-FLOOR, UNDERLAYMENT, C-C Plugged and C-D Plugged grades over 5/8" through 3/4" thick.
(3) See Section 3.8.1 for conditions under which stresses may be increased.

By permission of APA

## Conditions of Use

There are only two conditions of use which apply to plywood allowable stresses: wet and dry. If the equilibrium moisture content will be less than 16% in service, the dry use values should be used. As long as the plywood is not directly exposed to the weather, this dry condition can be assumed.

## Allowable Stress Table

Table 23 lists the allowable stresses and stiffnesses which should be used in plywood design.

## Duration of Load Factors

Plywood is a wood product, so its load capacity is time-dependent. The load duration factors of table 2 also apply to the allowable plywood stresses of table 23.

- **Width of Plywood Strip:** The allowable stresses of table 23 apply to plywood panels which are at least 24" wide. If the plywood is used in narrower strips, there is an increased possibility of a defect appearing in a critical section. Allowable stresses should be linearly decreased from full-strength at 24" wide to half-strength at 8" wide.

## D. Diaphragms and Shear Walls

Wind and seismic forces are the principal lateral forces a structure must resist. With minor modifications, building walls can act as shear walls, and the floors and roof can act as diaphragms to efficiently resist these lateral loads. Problems involving complex layout or critical diaphragm loading require analyses that are well beyond the scope of this book. Other sources, such as those listed as references for this book, should be consulted.

## Diaphragms

Floors and roofs can be designed to act as very deep, horizontal beams which carry the lateral forces applied to the walls between the floors and roof. These deep beams, or diaphragms, span shear walls and other structural elements carrying the lateral loads to the building foundation.

Just as with other beams, diaphragms are designed to resist the imposed shear and bending stresses. The shear stress is carried by the plywood decking, and is assumed to be uniformly distributed across the depth. The nailing schedule and panel splicing details required for a given design shear force are determined from table 24.

### Table 24
### Required Panel Details -- Diaphragms

**Recommended Shear in Pounds Per Foot for Horizontal Plywood Diaphragms for Wind or Seismic Loading**
(Plywood and framing assumed already designed for perpendicular loads)

| Plywood grade (c) | Common nail size | Minimum nail penetration in framing (in.) | Minimum nominal plywood thickness (in.) | Minimum nominal width of framing member (in.) | Blocked diaphragms | | | | Unblocked diaphragms | |
|---|---|---|---|---|---|---|---|---|---|---|
| | | | | | Nail spacing at diaphragm boundaries (all Cases and continuous panel edges parallel to load (in.) (Cases 3 & 4, and all panel edges Cases 5 & 6) (a) | | | | Nails spaced 6" max. at supported edges (a) | |
| | | | | | 6 | 4 | 2-1/2 | 2 | Case 1, where there are neither unblocked edges nor continuous panel joints parallel to load | All other configurations (Cases 2, 3, 4, 5 & 6) |
| | | | | | Nail spacing at other plywood panel edges (in.) (Cases 1, 2, 3 & 4) | | | | | |
| | | | | | 6 | 6 | 4 | 3 | | |
| STRUCTURAL I C-D INT-APA or STRUCTURAL I C-C EXT-APA | 6d | 1-1/4 | 5/16 | 2<br>3 | 185<br>210 | 250<br>280 | 375<br>420 | 420<br>475 | 165<br>185 | 125<br>140 |
| | 8d | 1-1/2 | 3/8 | 2<br>3 | 270<br>300 | 360<br>400 | 530<br>600 | 600<br>675 | 240<br>265 | 180<br>200 |
| | 10d | 1-5/8 | 1/2 | 2<br>3 | 320<br>360 | 425<br>480 | 640(b)<br>720 | 730(b)<br>820 | 285<br>320 | 215<br>240 |
| C-C EXT-APA, STRUCTURAL II C-D INT-APA, STRUCTURAL II C-C EXT-APA, C-D INT-APA Sheathing and other APA grades except Species Group 5. | 6d | 1-1/4 | 5/16 | 2<br>3 | 170<br>190 | 225<br>250 | 335<br>380 | 380<br>430 | 150<br>170 | 110<br>125 |
| | | | 3/8 | 2<br>3 | 185<br>210 | 250<br>280 | 375<br>420 | 420<br>475 | 165<br>185 | 125<br>140 |
| | 8d | 1-1/2 | 3/8 | 2<br>3 | 240<br>270 | 320<br>360 | 480<br>540 | 545<br>610 | 215<br>240 | 160<br>180 |
| | | | 1/2 | 2<br>3 | 270<br>300 | 360<br>400 | 530<br>600 | 600<br>675 | 240<br>265 | 180<br>200 |
| | 10d | 1-5/8 | 1/2 | 2<br>3 | 290<br>325 | 385<br>430 | 575(b)<br>650 | 655(b)<br>735 | 255<br>290 | 190<br>215 |
| | | | 5/8 | 2<br>3 | 320<br>360 | 425<br>480 | 640(b)<br>720 | 730(b)<br>820 | 285<br>320 | 215<br>240 |

(a) Space nails 12 in. on center along intermediate framing members.
(b) Reduce tabulated allowable shears 10% when boundary members provide less than 3-inch nominal nailing surface.
(c) All recommendations based on the use of APA grade-trademarked plywood.

Notes: Design for diaphragm stresses depends on direction of continuous panel joints with reference to load, not on direction of long dimensions of plywood sheet. Continuous framing may be in either direction for blocked diaphragms.

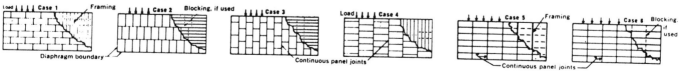

By permission of APA

In both tables 24 and 25, blocking the unsupported plywood edges between the principal framing members significantly increases the capacity.

The bending forces in diaphragms are usually resisted by the roof or floor perimeter framing, which act as the diaphragm chords -- comparable to the chords of a truss. The chord is sized to resist the calculated bending forces. Since diaphragms are commonly much longer than available or manageable lumber lengths, the chords must be spliced adequately.

Shear Walls

Building walls which are parallel to an applied lateral force can carry that force down to the foundation as short, deep cantilevers. Once the shear forces along the shear wall-diaphragm intersection are determined, table 25 can be used to determine the plywood thickness, panel layout, and nailing schedule required to provide the design capacity.

## Table 25
### Plywood Shear Wall Capacities
**Recommended Shear in Pounds Per Foot for Plywood Shear Walls** for Wind or Seismic Loading (a)

| Plywood grade | Minimum nominal plywood thickness (in.) | Minimum nail penetration in framing (in.) | Plywood applied direct to framing | | | | | Plywood applied over 1/2" gypsum sheathing | | | | |
|---|---|---|---|---|---|---|---|---|---|---|---|---|
| | | | Nail size (common or galvanized box) | Nail spacing at plywood panel edges (in.) | | | | Nail size (common or galvanized box) | Nail spacing at plywood panel edges (in.) | | | |
| | | | | 6 | 4 | 2½ | 2 | | 6 | 4 | 2½ | 2 |
| STRUCTURAL I C-D INT-APA, or STRUCTURAL I C-C EXT-APA | 5/16 or 1/4(b) | 1-1/4 | 6d | 200 | 300 | 450 | 510 | 8d | 200 | 300 | 450 | 510 |
| | 3/8 | 1-1/2 | 8d | 280 | 430 | 640 | 730 | 10d | 280 | 430 | 640(d) | 730(d) |
| | 1/2 | 1-5/8 | 10d | 340 | 510 | 770(d) | 870(d) | — | — | — | — | — |
| C-C EXT-APA, STRUCTURAL II C-D INT-APA, STRUCTURAL II C-C EXT-APA, C-D INT-APA Sheathing APA panel siding(e) and other APA grades except species Group 5. | 5/16 or 1/4 (b) | 1-1/4 | 6d | 180 | 270 | 400 | 450 | 8d | 180 | 270 | 400 | 450 |
| | 3/8 | 1-1/2 | 8d | 260 | 380 | 570 | 640 | 10d | 260 | 380 | 570(d) | 640(d) |
| | 1/2 | 1-5/8 | 10d | 310 | 460 | 690(d) | 770(d) | — | — | — | — | — |
| | | | Nail size (galvanized casing) | | | | | Nail size (galvanized casing) | | | | |
| APA panel siding(e) applied with casing nails(c) | 5/16 (b) | 1-1/4 | 6d | 140 | 210 | 320 | 360 | 8d | 140 | 210 | 320 | 360 |
| | 3/8 | 1-1/2 | 8d | 160 | 240 | 360 | 410 | 10d | 160 | 240 | 360 | 410 |

(a) All panel edges backed with 2-inch nominal or wider framing. Plywood installed either horizontally or vertically. Space nails at 12 in. on center along intermediate framing members.
(b) Minimum recommended when applied direct to framing as exterior siding is 3/8" or 303-16 o.c.
(c) Except Group 5 species.
(d) Reduce tabulated shears 10% when boundary members provide less than 3-inch nominal nailing surface.
(e) 303-16 o.c. plywood may be 5/16", 3/8" or thicker. Thickness at point of nailing on panel edges governs shear values.

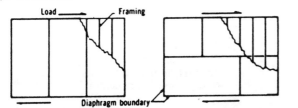
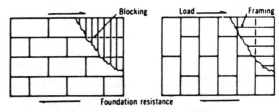

By permission of APA

## Design Method -- Shear Walls and Diaphragms

The basic design procedure is to determine the applied loads and detail the respective elements to carry the loads.

Step 1:
  Calculate applied loads as shears (pounds/foot) along the supported edge of diaphragms or loaded edge of shear walls.

Step 2:
  Determine panel layout, plywood thickness, and nailing schedule from table 24 or 25.

Step 3:
  Determine diaphragm chord size and detail splices.

Step 4:
  Check deflections by comparing length-width ratios to allowable ones.

Step 5:
  Detail connections between elements and to the foundation.

## Example 17

Determine the design shear on the diaphragms and shear walls of the building shown. Design and detail the roof as a diaphragm and the first floor interior wall as a shear wall. Assume a wind load of 25 psf and consider only wind against the long side of the building. Use KD Southern pine for the framing.

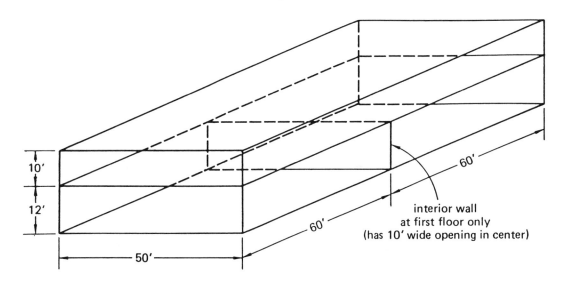

Determine the wind loads applied to the long edges of the roof and floor from the wall which frames between them. Assume a tributary area distribution of the uniform load.

Load on the 120 foot length of the roof = (25) psf (10) ft/(2)
= 125 plf

Load on the 120 foot length of the second floor = (25) [10 + 12] ft/2
= 275 plf

Note that the wind load on the lower 6 feet of the wall is carried directly to the foundation and represents a relatively small shear force.

Determine the shears along the supported edges of the diaphragms (where they are connected to the shear walls.)

Roof-to-end wall connection shear = (125) plf [(120) ft/2]/(50) ft
= 150 plf

Floor-to-end wall shear = (275) plf [(60) ft/2]/(50) ft
= 165 plf

Floor-to-centerline wall = (5/8) (120) ft (275) plf/(40) ft = 516 plf

Note that the last shear calculation includes the 5/8 factor applicable to the center support of a beam continuous over three supports, even though the endwall calculation is based on a simple tributary area calculation. This assumed behavior is conservative and indicative of the simplified analyses justified in load distribution calculations.

To design the roof as a diaphragm, first find an adequate configuration in table 24. Since blocking can be expensive, first try to find an option with unblocked edges. The lower group of plywood grades is the cheaper, so also try to find a layout in that section of table 23. Several layouts meet the design requirement for 150 plf. In practice, the choice might be influenced by snow loads, local practice, or other architectural considerations. For this case, assume 2 inch nominal framing will be adequate and that 3/8 inch plywood will be adequate for the snow load. A 160 plf capacity is provided by the following configuration:

> C-C EXT-APA, 3/8" plywood, 8d nails on 6" centers along the endwall and on 12" centers along the other framing. The edges can be unblocked, and 2" nominal framing members are adequate. The panel layout can be Case 2, 3, 4, 5, or 6.

The chord force is determined by finding the maximum bending moment and dividing by the diaphragm depth -- the "moment arm" of the chord forces.

The maximum moment is

$$M_{max} = wL^2/8 = (125) \text{ plf } (120)^2 \text{ ft}^2/8 = 225,000 \text{ ft-lb}$$

The bending force in the chords is, therefore,

$$M_{max}/d = (225,000) \text{ ft-lb}/(50) \text{ ft} = 4500 \text{ lb}$$

Since one chord will be in tension and the other in compression, the smaller of the two allowable stresses will control the chord size. Appendix 2 gives $F_c$ = 1200 psi and $F_t$ = 675 psi, so that the tension chord controls the design size. Furthermore, footnote 3 specifies a 40% reduction in $F_t$ for an assumed 10" or deeper member. The required chord area is, therefore,

$$(4500) \text{ lb}/(675) \text{ psi } (.60)(1.33) = 8.35 \text{ in}^2$$

1.33 is the load duration factor for wind loads. Assuming a doubled 2" nominal chord, splices will have a 1.5" wide continuous member. The required depth of the chord is

$$(8.35) \text{ in}^2/(1.5) \text{ in} = 5.57 \text{ in}$$

Even after removing some of the chord section for the splice fasteners, any chord of doubled 2x8's or larger will be adequate. The splices should be designed to handle the maximum chord force.

The diaphragm is only as good as the connection which supports it at the shear walls. This connection should have a design capacity of 160 plf, the capacity of the diaphragm--not the 150 plf design load. There are several ways to achieve this capacity. One efficient way is to have the plywood wall sheathing overlap the roof perimeter framing.

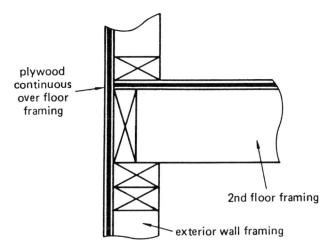

**Roof-Shear wall Connection Detail**

The Uniform Building Code (and most other codes) specifies a maximum length-width ratio of 4:1 for horizontal diaphragms sheathed with plywood. This ratio is intended to eliminate excessive deflection. The ratio in this example is (120) ft/(50) ft = 2.4:1, so the criterion is satisfied.

The interior wall has a shear load of 515 plf, applied on the top by the second floor and transmitted down to the foundation. Table 25 distinguishes between walls with the plywood applied directly to the framing and plywood applied over 1/2" sheetrock, intended as fire walls. Assuming this is not a fire wall, a 570 plf load can be carried in a shear wall with the following specifications:

> 3/8" C-D INT-APA Sheathing, with 8d nails 2 1/2" on center at the boundary framing members and 12" on center at all other framing. Any of the panel layouts illustrated at the bottom of table 25 are acceptable.

The chords of shear walls are the vertical framing members at the corners and around openings. Since standard framing practice results in triple members at corners and doubled members at openings, the chords at the openings control. The load at the top of each half of the shear wall is

(515) plf (20) ft = 10,030 lb

This force creates a bending moment that is maximum at the bottom of the wall, where it is equal to

(10,030) lb (12) ft = 123,600 ft-lb

12 feet is the height of the wall. This causes a bending force in the vertical chords equal to the applied moment divided by the chord separation, or shear wall length:

Chord force = (123,600) ft-lb/(20) ft = 6,180 lb

Using the allowable tension stress calculated for the roof diaphragm, this chord force requires an area of

(6,180) lb/(.60)(1.33)(675) psi = 11.47 in$^2$

Assuming the verticals will be short enough to be one piece, the full 3" width will be available. This requires a vertical framing member depth of

(11.47) in$^2$/(3) in = 3.82 in

This is wider than the 3.5 inches of a 2x4, so a 2x6 wall is required. If 2x6's cannot be used, a higher grade framing in this wall will be necessary.

The wall should be connected to the foundation to resist both the shear load of 570 plf and the uplift load in the chords of 6,180 pounds at each corner and side of the opening. Anchor bolts at closer than standard spacings or steel fittings can be used to provide this connection capacity.

The height-width ratio for this shear wall is (12) ft/(40) ft, or 0.30:1, less than the 1:1 ratio for which deflections should be calculated. Both 20' sections of the interior wall are taken into account. This issue of openings in plywood diaphragms and shear walls is very complex and is at the cutting edge of the research in the field. This text has necessarily treated this subject lightly.

### E. Plywood-Lumber Built-Up Beams

With its alternating grain directions, plywood is very strong in shear. Its panel configuration also tends to make plywood a good material for webs of built-up beams. Lumber, with all its fibers oriented in the same direction, can efficiently resist the axial forces found in the flanges of built-up beams. For spans longer or loads heavier than dimension lumber beams can handle, plywood-lumber beams can be cheaper and easier to obtain than glue-laminated members.

There are many design variables when designing a plywood-lumber beam. The beam depth and configuration can be changed. The size, number, species, and grade of the lumber flanges are variable. Finally, the webs can be any number of any thickness and type plywood.

With all these available variables, there are two basic design methodologies. Several agencies publish tables of plywood-lumber beam capacities, spans, and details. The other choice is to pick a trial section based on experience and available material. The shear stresses, bending stresses, and deflections should then be checked. If the trial section is adequate without being unreasonably understressed, the design is completed by detailing the connections.

Trying to achieve an optimal built-up beam design, with all the design variables, could be an endless task. Unless the beams are to be mass produced, it is unnecessary to spend a lot of time optimizing design. A more important consideration is material availability.

The following design procedure is based on Supplement 2 of the Plywood Design Specifications, published by the American Plywood Association.

This publication contains much valuable information on fabrication detailing, unsymmetrical sections, and detailing. Its use is highly recommended.

Design Considerations

The two basic configuration layouts used in plywood-lumber beams are the I beam and box beam.

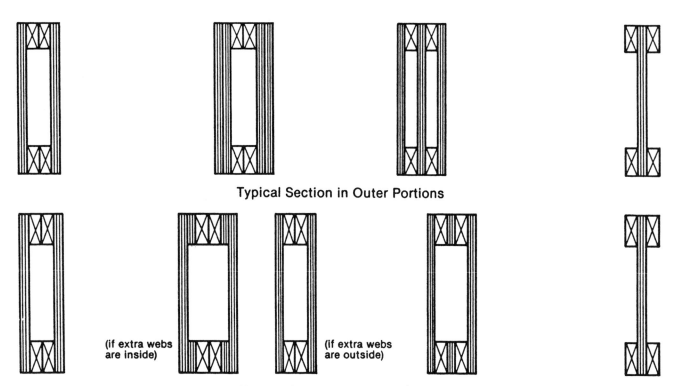

Figure 47
Plywood-Lumber Beam Cross-Sections
By permission of APA

The allowable stresses for the lumber flanges are found in the National Design Specifications, excerpted in appendix 2. Table 23 provides the allowable stresses in the plywood webs.

Deflections are calculated through standard mechanics procedures, being limited to the same values as solid beams. Since these beams tend to be long and/or heavily loaded, and are fabricated instead of sawn, camber is a design consideration. The recommended camber is 1.5 times the dead load deflection.

## Table 26
### Preliminary Capacities of Plywood-Lumber Beam Cross-Sections
By permission of APA

Plywood webs, butt joints staggered 24" minimum, spliced per PDS Section 5.6.3.2

Continuous lumber flanges (no butt joints), resurfaced for gluing per Part 1, Section 4.1.2

| Depth, Flange | Max. Moment,[1][2] M (ft-lb) | | | Max. Shear,[1] V (lb) |
|---|---|---|---|---|
| | $M_{flange}$ | $M_{web}$[3][4] | $M_{total}$ | $V_{horizontal}$[5][6] |
| 12"  1-2x4 | 2375 | 391 | 2766 | 1340 |
| 2-2x4 | 4751 | 391 | 5142 | 1358 |
| 3-2x4 | 7126 | 391 | 7517 | 1365 |
| 16"  1-2x4 | 3771 | 706 | 4477 | 1884 |
| 2-2x4 | 7543 | 706 | 8249 | 1928 |
| 3-2x4 | 11314 | 706 | 12020 | 1947 |
| 1-2x6 | 4510 | 706 | 5216 | 1741 |
| 2-2x6 | 9019 | 706 | 9725 | 1755 |
| 3-2x6 | 13529 | 706 | 14235 | 1761 |
| 20"  1-2x4 | 5217 | 1114 | 6331 | 2421 |
| 2-2x4 | 10434 | 1114 | 11548 | 2497 |
| 3-2x4 | 15652 | 1114 | 16766 | 2530 |
| 1-2x6 | 6646 | 1114 | 7760 | 2283 |
| 2-2x6 | 13291 | 1114 | 14405 | 2317 |
| 3-2x6 | 19937 | 1114 | 21051 | 2331 |
| 1-2x8 | 7193 | 1114 | 8307 | 2163 |
| 2-2x8 | 14386 | 1114 | 15500 | 2176 |
| 3-2x8 | 21580 | 1114 | 22694 | 2181 |
| 24"  1-2x4 | 6554 | 1598 | 8152 | 2935 |
| 2-2x4 | 13108 | 1598 | 14706 | 3046 |
| 3-2x4 | 19662 | 1598 | 21260 | 3096 |
| 1-2x6 | 8761 | 1598 | 10359 | 2815 |
| 2-2x6 | 17522 | 1598 | 19120 | 2876 |
| 3-2x6 | 26283 | 1598 | 27881 | 2901 |
| 1-2x8 | 9850 | 1598 | 11448 | 2687 |
| 2-2x8 | 19699 | 1598 | 21297 | 2719 |
| 3-2x8 | 29549 | 1598 | 31147 | 2732 |
| 1-2x10 | 10411 | 1598 | 12009 | 2559 |
| 2-2x10 | 20822 | 1598 | 22420 | 2570 |
| 3-2x10 | 31234 | 1598 | 32832 | 2575 |
| 30"  2-2x4 | 17486 | 2518 | 20004 | 3873 |
| 3-2x4 | 26229 | 2518 | 28747 | 3951 |
| 4-2x4 | 34972 | 2518 | 37490 | 3997 |
| 2-2x6 | 24358 | 2518 | 26876 | 3730 |
| 3-2x6 | 36537 | 2518 | 39055 | 3776 |
| 4-2x6 | 48716 | 2518 | 51234 | 3801 |
| 2-2x8 | 28417 | 2518 | 30935 | 3572 |
| 3-2x8 | 42626 | 2518 | 45144 | 3601 |
| 4-2x8 | 56835 | 2518 | 59353 | 3616 |
| 2-2x10 | 31270 | 2518 | 33788 | 3396 |
| 3-2x10 | 46904 | 2518 | 49422 | 3411 |
| 4-2x10 | 62539 | 2518 | 65057 | 3419 |
| 2-2x12 | 32693 | 2518 | 35211 | 3245 |
| 3-2x12 | 49039 | 2518 | 51557 | 3252 |
| 4-2x12 | 65385 | 2518 | 67903 | 3255 |
| 36"  2-2x4 | 21894 | 3647 | 25541 | 4683 |
| 3-2x4 | 32842 | 3647 | 36489 | 4793 |
| 4-2x4 | 43789 | 3647 | 47436 | 4858 |
| 2-2x6 | 31324 | 3647 | 34971 | 4576 |
| 3-2x6 | 46985 | 3647 | 50632 | 4646 |
| 4-2x6 | 62647 | 3647 | 66294 | 4685 |
| 2-2x8 | 37442 | 3647 | 41089 | 4429 |
| 3-2x8 | 56163 | 3647 | 59810 | 4476 |
| 4-2x8 | 74884 | 3647 | 78531 | 4502 |
| 2-2x10 | 42368 | 3647 | 46015 | 4246 |
| 3-2x10 | 63553 | 3647 | 67200 | 4276 |
| 4-2x10 | 84737 | 3647 | 88384 | 4293 |
| 2-2x12 | 45487 | 3647 | 49134 | 4072 |
| 3-2x12 | 68231 | 3647 | 71878 | 4090 |
| 4-2x12 | 90975 | 3647 | 94622 | 4099 |
| 42"  2-2x6 | 38362 | 4983 | 43345 | 5411 |
| 3-2x6 | 57543 | 4983 | 62526 | 5508 |
| 4-2x6 | 76725 | 4983 | 81708 | 5563 |
| 2-2x8 | 46640 | 4983 | 51623 | 5280 |
| 3-2x8 | 69960 | 4983 | 74943 | 5350 |
| 4-2x8 | 93280 | 4983 | 98263 | 5388 |
| 2-2x10 | 53836 | 4983 | 58819 | 5103 |
| 3-2x10 | 80754 | 4983 | 85737 | 5150 |
| 4-2x10 | 107672 | 4983 | 112655 | 5176 |
| 2-2x12 | 58956 | 4983 | 63939 | 4921 |
| 3-2x12 | 88434 | 4983 | 93417 | 4953 |
| 4-2x12 | 117912 | 4983 | 122895 | 4970 |
| 48"  2-2x6 | 45446 | 6529 | 51975 | 6235 |
| 3-2x6 | 68170 | 6529 | 74699 | 6361 |
| 4-2x6 | 90893 | 6529 | 97422 | 6434 |
| 2-2x8 | 55946 | 6529 | 62475 | 6125 |
| 3-2x8 | 83919 | 6529 | 90448 | 6219 |
| 4-2x8 | 111892 | 6529 | 118421 | 6272 |
| 2-2x10 | 65533 | 6529 | 72062 | 5958 |
| 3-2x10 | 98300 | 6529 | 104829 | 6026 |
| 4-2x10 | 131066 | 6529 | 137595 | 6063 |
| 2-2x12 | 72843 | 6529 | 79372 | 5777 |
| 3-2x12 | 109265 | 6529 | 115794 | 5825 |
| 4-2x12 | 145686 | 6529 | 152215 | 5851 |

## Trial Section

To determine a trial section, choose the beam depth as 1/8 to 1/12 of the span, with a depth which makes efficient use of the 4 foot module of commercially available plywood. Table 26 gives very approximate values for the shear and moment capacities of various cross-sections. These are only preliminary cross-sections, and they must be checked against the actual design and allowable stresses.

## Lumber Flanges

Since the lumber flanges are primarily loaded axially by the bending stresses, $F_c$ and $F_t$ are the considered allowable stresses. If the beam is symmetrical, the smaller of the two allowable axial stresses will control. The bending stress equation (22) is used with the net moment of inertia, $I_n$, which neglects those longitudinal fibers in the webs and flanges which are interrupted by butt splices.

When calculating $I_n$, three reductions in the lumber flange area must be considered. The first is due to the resurfacing required for a competent glue joint at the web/flange connection. Assume this is a 1/8" reduction in width across the beam cross-section. The second area reduction is due to resurfacing the depth of the assembled beam, intended to smooth out irregularities in the fit along the top and bottom of the beam. Assume this reduction in beam and flange depth to be 3/8" for beams less than 24" deep, and 1/2" for beams 24" deep or deeper.

The last flange area reduction results from butt splicing the lumber to achieve the required beam length. The reduction is a function of the butt joint spacings. The butt jointed flange members are neglected in any $I_n$ calculation. If two flange pieces are butt jointed at a spacing less than 10 times the thickness of the flange pieces, both flange pieces are neglected at the more critical of the two locations. Unless the splices are spaced more than 50 times the flange member width, the unjointed flange members also have their areas reduced according to the factor of table 27.

## Table 27
### Effective Area of Unspliced Flange Members

| Butt-Joint Spacing (t = lamination thickness) | Effective Laminae Area |
|---|---|
| 30t | 90% |
| 20t | 80% |
| 10t | 60% |

When checking flange tension stresses, the jointed members are neglected, the unjointed members reduced with table 27, and the allowable stress is further reduced by 20%.

The plywood webs contribute some bending resistance. This is accounted for in the $I_n$ calculation by including only the plywood plies parallel to the longitudinal beam axis. Butt-spliced web members must be neglected in this calculation, unless they are spliced full-depth.

### Plywood Webs

The major stress in the plywood web is through-the-panel shear stress. When calculating the maximum value of the shear stress (at the neutral axis), equation 9 is used. In this calculation, however, the Q and I section properties include all longitudinal fibers, regardless of butt splicing. The thickness in equation 9 is the sum of all shear thicknesses of plywood present at the section.

### Flange to Web Connection

The shear connection between the web(s) and flange(s) cause rolling shear in the plywood. The shear stress is evaluated at the glue line with equation 9. This stress is compared with an allowable rolling shear from table 23. The table value should be reduced by 50% to account for stress concentrations that arise at the connection.

### Deflections

The APA Supplement 2 contains a refined way to calculate the separate bending and shear components of deflection in plywood-lumber beams. A more approximate method is to calculate the bending component with standard deflection equations, and then increase it to account for shear deflection. The magnitude of the increase is a function of the span-to-depth ratio, as indicated in table 28.

**Table 28**
**Bending Deflection Increase to Account for Shear**

| Span/Depth | Increase |
|---|---|
| 10 | 50% |
| 15 | 20% |
| 20 | 0% |

The values of table 28 may be linearly interpolated for actual span-to-depth ratios.

### Details

Any splices must be detailed to transmit the resultant design forces. In addition, stiffeners are added wherever a point load or support reaction is applied to the beam. These stiffeners fit vertically between the flanges to reinforce the web and distribute the point load. They are sized to not crush the flanges at the bearing points, and to not induce rolling shear failure in the web, as the load is transferred from flange to web.

### Lateral Stability

Plywood-lumber beams can be long and subject to lateral buckling. The American Plywood Association suggests considering the ratio of cross-sectional stiffnesses about the vertical and horizontal axes as a measure of the beam's tendency to buckle laterally. The moments of inertia include all longitudinal fibers, regardless of splicing. The bracing recommendations, as a function of that ratio, are found in table 29.

**Table 29**
**Lateral Bracing Required for Plywood-Lumber Beams**

| $I_x/I_y$ | Lateral Bracing Required |
|---|---|
| $\leq 5$ | None |
| 5 - 10 | Ends held at bottom at supports |
| 10 - 20 | Top and bottom held at ends |
| 20 - 30 | Top (or bottom) edge held |
| 30 - 40 | Bridging or equivalent on 8 foot centers, or less |
| $\geq 40$ | Compression flange held by well-fastened sheathing, or equal |

# Example 18

Given: A 24" deep (nominal) plywood-lumber beam with a 25' clear span. The webs are 5/8" 32/16 APA Rated Sheathing with exterior glue. The webs are not spliced at the butt joints. The flanges are 3 DF-L, dense #1, 2"x6"'s, which are butt-jointed on 48" centers. The standard resurfacing has been done. The beam is subjected to normal duration and dry loading. Deflection is limited to span/360. What is the beam's uniform load capacity as limited by (a) flange stresses, (b) web stresses, (c) web/flange connection stresses, and (d) deflection.

The first step is to evaluate the section properties. Butt-jointed members are considered as intermittently effective in the net section properties. Once the section properties are known, the design equations are solved for the allowable uniform load.

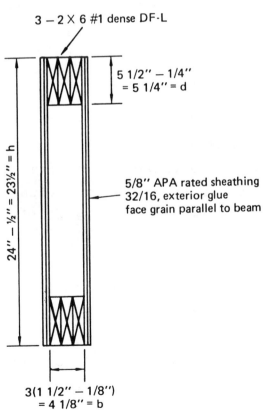

The members are resurfaced during beam fabrication. The flange members are sanded 1/8" thinner for better glueing. After assembly, the beam's depth is resurfaced and reduced by 1/2". This depth reduction is assumed to reduce the depth of each flange by 1/4".

## Section Properties

Moment of Inertia, I (total and net)

$$I_{flanges} = b/12\,[h^3 - (h-2d)^3]$$
$$= [(2)(.90)(1.5"-1/8")]/(12)\,[(23.5)^3 - \{(23.5)-(2)(5.25)\}^3]$$
$$= 2{,}224 \text{ in}^4$$

This is the net moment of inertia for the flanges, where one of the laminae is neglected (leaving 2 pieces), and the remaining members are only 90% effective. The total moment of inertia includes all parallel fibers, regardless of butt splicing.

$$I_{flanges} = (2{,}224) \text{ in}^4 \, (3 \text{ pieces}/2 \text{ pieces})\,/\,(0.9)$$
$$= 3{,}707 \text{ in}^4$$

$$I_{webs} = t_{\parallel}\,h^3/12 = \frac{2.951 \text{ in}^2/\text{ft}\,(23.5)^3 \text{in}^3}{12 \text{ in/ft}\,(12)}$$

$$= 266 \text{ in}^4, \text{ (per web)}$$

Only the plywood fibers which are parallel to the beam axis contribute to section properties. The effective thickness, $t_{\parallel}$, is derived from the area of parallel fibers per foot of width, 2.951 in$^2$/ft. Figure 46 gives the group 3 species, and table 21 gives the unsanded finish and S-1 stress rating.

The composite section properties are a combination of the plywood webs and the lumber flanges.

$$I_{net} = 2{,}224 + 266 = 2{,}490 \text{ in}^4 \text{ (only one web is considered)}$$
$$I_{total} = 3{,}706 + (2)(266) = 4{,}238 \text{ in}^4 \text{ (all parallel fibers count)}$$

$$Q_{flanges} = bd(h/2 - d/2)$$
$$= (4.125")(5.25")(23.5"/2 - 5.25"/2)$$
$$= 197.6 \text{ in}^3$$

$$Q_{webs} = (t_\parallel)(h/2)(h/4)(\text{\# of webs})$$
$$= (2.951 \text{ in}^2/\text{ft})(1 \text{ ft}/12 \text{ in})(23.5 \text{ in}/2)(23.5 \text{ in}/4)(2)$$
$$= 34.0 \text{ in}^3$$

$$Q_{total} = Q_{webs} + Q_{flanges} = 197.6 + 34.0 = 231.6 \text{ in}^3$$

Note that the only Q used is the total value. Therefore, a net value is not calculated. In both I and Q calculations, the web contribution is much smaller than that of the flanges. This is one justification for not correcting for the different stiffnesses of the lumber flanges and plywood webs before combining them.

(a) Allowable load, as limited by bending stresses in the flanges:

The allowable tensile stress for the lumber flanges is 1200 psi, found in appendix 2. This controls over the allowable stress in the compression flange, $F_c$, 1450 psi. $F_t$ is further reduced by 20% at the butt-jointed sections.

$$\text{The allowable moment} = (F_t)(I_n)/(.5)(h)$$
$$= (0.8)(1200) \text{ psi } (2,492) \text{ in}^4/(.5)(23.5) \text{ in}$$
$$= 203,602 \text{ in-lb} = 16,967 \text{ ft-lb}$$

$$\text{The allowable load} = (M_{all})(8)/(\text{span})^2$$
$$= (16,967) \text{ ft-lb } (8)/(25)^2 \text{ ft}^2$$
$$= 217 \text{ plf}$$

(b) Allowable load, limited by shear stress in web:

The Plywood Design Specifications allow a 33% increase in allowable shear through the thickness of plywood if the plywood panel is rigidly glued to continuous framing around its edges. Table 23 gives $F_v$ as 140 psi, for S-1, species 3, dry use. For this application,

$$F_v = (1.33)(140) = 186.7 \text{ psi}.$$

The allowable horizontal shear is,

$$V_h = F_v I_t t_s / Q$$

$$V_h = [(186.7) \text{ psi} (4{,}238) \text{ in}^4 (2 \text{ webs}) (0.336) \text{ in/web}] / (231.6) \text{ in}^3$$
$$= 2{,}296 \text{ lb}$$

Neglecting the uniform load within a beam depth of the supports, the allowable uniform load for this allowable horizontal shear is

$$w_{all} = 2V/(L - 2h) = (2)(2{,}295) \text{ lb}/[25 - \{(2)(23.5)/(12)\}]$$
$$= 218 \text{ plf}$$

(c) Allowable load, limited by rolling shear at the web/flange connection:

The rolling shear at the glue line between web and flange, $V_s$, is

$$V_s = 2F_s d I_t / Q_{flanges} = \frac{(2)[(53) \text{ psi}/(2)] (5.25) \text{ in} (4{,}238) \text{ in}^4}{(197.6) \text{ in}^3}$$
$$= 5{,}968 \text{ lb}$$

Note that the allowable rolling shear stress was halved because of the shear stress concentrations at the edge of the panel. This allowable shear translates into the following allowable uniform load.

$$w_{all} = (218) \text{ plf} \frac{(5{,}968) \text{ lb}}{(2{,}295) \text{ lb}} = 567 \text{ plf}$$

(d) Allowable load, limited by deflection:

A simple method of calculating deflections in plywood-lumber beams is to increase calculated bending deflections by a factor to account for the neglected, but significant, shear deflections. Table 28 gives this factor as a function of the span-to-depth ratio.

$$\text{span/depth} = (25) \text{ ft}/[(23.5) \text{ in}/(12) \text{ in/ft}] = 12.8$$

Interpolating between ratios of 10 and 15 yields a factor of 1.33.

The allowable deflection, span/360, is equal to (25)(12)/(360) = 0.83" Use the maximum bending deflection equation for uniform load on simple spans to calculate the allowable uniform load.

$$w_{all} = \frac{(.83) \text{ in } (384)(1.03)(1,900,000) \text{ psi } (4,238) \text{ in}^4}{(1.33)(5)(25)^4 \text{ ft}^4 (1,728) \text{ in}^3/\text{ft}^3}$$

$$= 589 \text{ plf}$$

Note that a 3% increase in the lumber flange E was included, because the extra deflection due to shear is accounted for with the 33% increase. The built-in shear deflection allowance in the tabled E value of appendix 2 is redundant and, therefore, eliminated.

The uniform load capacity of the plywood-lumber beam is limited by bending stresses to 217 plf, and by shear in the plywood webs to 218 plf. There are several ways to increase the capacity: use thicker plywood, use a higher grade plywood -- such as structural 1, with its species 1 allowable stresses, or simply use more webs of the same plywood at the beam ends where shear streses are maximum. The flanges could also be increased in number or size, or a higher stress species or grade could be specified.

**RESERVED FOR FUTURE USE**

## 11. SAMPLE PROBLEMS

The first five sample problems are similar to problems from past professional engineering exams. They represent, therefore, the level of complexity that could be expected on future exams. The proposed solutions are the author's own, and they are not necessarily the answers used by the grader for the problem in the year it was given. The other problems approximate the same level of complexity.

## SAMPLE PROBLEM 1

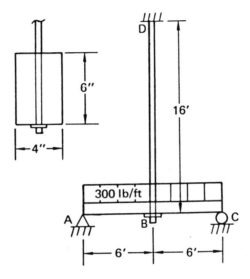

A beam supported by non-yielding supports at the ends and a steel rod at midspan has the dimensions and loads shown.

The beam is of timber, 4 inches wide by 6 inches deep, having a modulus of elasticity of $1.5 \times 10^6$ psi.

The rod is 3/8 inch in diameter and $E = 30 \times 10^6$ psi. Assume the rod passes through a 1/2 inch diameter hole in the beam.

Find:
  (a) The reactions at A, B, and C.
  (b) The bending moment at B.
  (c) The maximum positive moment in the beam.
  (d) The maximum flexure stress in the beam.

# Timber Design

This beam is indeterminate to the first degree. A compatibility equation for the beam deflection at the centerline will provide the information required to analyze the entire beam.

The compatibility equation is

$Y_{\text{beam due to loads, without rod}} + Y_{\text{beam; rod force}} = $ rod elongation

Solve for the three terms of this equation, in terms of the unknown rod force.

$Y_{\text{beam due to loads, without rod}}$ (beam will deflect downward)

$$= 5wL^4/384EI = \frac{(5)(300)\text{lb/ft}(12)^4\text{ft}^4(1728)\text{in}^3/\text{ft}^3}{(384)(1.5 \times 10^6)\text{lb/in}^2(72)\text{in}^4} = 1.30 \text{ in}$$

This equation is found in any structural handbook or text. The 72 in$^4$ is the moment of inertia, $bd^3/12$, of the full-size 4"x6" cross-section.

$Y_{\text{beam; rod force}}$ (tension in rod will deflect beam upward)

$$= PL^3/48EI = \frac{P_{\text{rod}}(12)^3\text{ft}^3(1728)\text{in}^3/\text{ft}^3}{(48)(1.5 \times 10^6)\text{lb/in}^2(72)\text{in}^4} = 5.76 \times 10^{-4} (P_{\text{rod}}) \text{ in}$$

Rod elongation (tension in rod corresponds to downward deflection)

$$= PL/AE = \frac{P_{\text{rod}}(16)\text{ft}(12)\text{in/ft}}{\pi[(3/8)/2]^2 \text{in}^2(30 \times 10^6)\text{lb/in}^2} = 5.79 \times 10^{-5} (P_{\text{rod}})$$

Inserting these values into the compatibility equation:

$1.30 \text{ in} - (5.76 \times 10^{-4})(P_{\text{rod}}) = (5.79 \times 10^{-5})(P_{\text{rod}})$

Solving this equation for $P_{\text{rod}}$,

$P_{\text{rod}} = (1.30)/(5.79 \times 10^{-5} + 5.76 \times 10^{-4}) = 2050$ pounds.

# Timber Design

Before going any further, it is wise to check the reasonableness of this answer. One easy way to check it is to compare it with statically determinate limits on the rod force. If the rod were infinitely stiff, the problem would be a simple two-span, continuous beam. The rod force under those conditions would be

$$(5/8)(12) \text{ ft } (300) \text{ lb/ft} = 2250 \text{ lb}$$

Since the rod is flexible, the load should be, and is, less than this upper bound. The lower limit on the rod force is zero at no stiffness at all, but a more meaningful lower bound is determined by assuming the beam is very flexible compared with the rod. This is an assumption of two simple spans, giving a rod force of

$$(1/2)(12) \text{ ft } (300) \text{ lb/ft} = 1800 \text{ lb}$$

The indeterminate solution for the rod force falls between these two bounds, and therefore seems reasonable.

The elongation of the rod, equal to the beam settlement, is

$$(2050) \text{ lb } (5.76 \times 10^{-5}) \text{ in/lb} = 0.12 \text{ in}$$

Now that the indeterminate beam has been "solved," use statics to find the rest of the quantities.

a) Symmetry shows $R_A = R_C = [(300) \text{ lb/ft } (12) \text{ ft} - (2050) \text{ lb}]/2$
    $= 775 \text{ lb}$

b) Bending moment at B

$M_B = [(775) \text{ lb } (6) \text{ ft} - (300) \text{ lb/ft } (6)^2 \text{ ft}^2/(2)] (12) \text{ in/ft}$
  $= -9,000 \text{ in-lb}$, compression on bottom

c) The maximum positive bending moment occurs where the shear is zero.

$V = 0$ at: $(775) \text{ lb}/(300) \text{ lb/ft} = 2.58 \text{ ft}$ from either end of the beam

$$M_{2.58'} = [(775)\text{lb}(2.58)\text{ft} - \{(300)\text{lb/ft}(2.58)^2\text{ft}^2/(2)\}](12)\text{in/ft}$$
$$= 12,000 \text{ in-lb}$$

d) The maximum bending stress will either be at 2.58' from the end, or at the center -- when the net section at the hole is considered.

At the maximum postive moment

$$S = bd^2/6 = (4)(6)^2/(6) = 24 \text{ in}^3$$

$$f_b = M/S = (12,000) \text{ in-lb}/(24) \text{ in}^3 = 500 \text{ psi}$$

At the centerline the hole is removed in the section modulus calculation.

$$S = (4 - 1/2) \text{ in } (6)^2 \text{ in}^2 /(6) = 21.0 \text{ in}^3,$$

$$f_b = (9,000) \text{ in-lb}/(21) \text{ in}^3 = 429 \text{ psi}$$

The maximum bending stress occurs 2.58 feet from either end, and is equal to 500 psi.


## SAMPLE PROBLEM 2

In order to gain additional clear floor area for use in a building, it is planned to remove the center post under the 4" x 8" (nominal dimensions) beam. A king post truss will be formed, as indicated in the second figure, making the beam continuous over the strut.

Calculate the stresses in the members of the trussed beam, and determine if they are within the allowable stresses indicated.

```
Data:     Timber:   Allowable "f" - 1400 psi
                              E   - 1,600,000 psi
          Steel:    Allowable "f" - 20,000 psi
                              E   - 30,000,000 psi
```

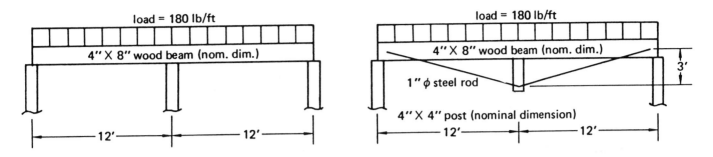

The "trussed beam" is indeterminate to the first degree. The king post truss layout was very popular in mill buildings in the 19th century. Handbook solutions of that era do exist, but will probably be more trouble to find than simply solving the problem.

This problem is similar to sample problem 1. A single compatibility equation for the beam deflection at the centerline will provide the information required to analyze the entire beam. The only real difference is determining the "elasticity" of the support provided by the king post. Solve for the deflection terms.

$Y_{beam}$; loaded, without the king post (downward)

$$= \frac{5wL^4}{384EI} = \frac{5(180) \text{lb/ft} (24)^4 \text{ft}^4 (1728) \text{in}^3/\text{ft}^3}{384(1.6 \times 10^6) \text{lb/in}^2 (111.148) \text{in}^4} = 7.56 \text{ in}$$

The 111.148 in$^4$ moment of inertia is found in Appendix 5.

$Y_{beam}$; due to an upward point load at center (upward)

$$= PL^3/48EI = \frac{(P_{king\ post})(24)^3 \text{ft}^3 (1728) \text{in}^3/\text{ft}^3}{48(1.6 \times 10^6) \text{lb/in}^2 (111.148) \text{in}^3}$$

$$= 2.80 \times 10^{-3} (P_{king\ post})$$

To establish the flexibility of the king post, place an imaginary one pound load on top of the post and determine the displacement with the virtual work method.

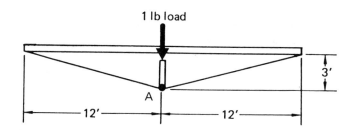

A truss analysis (starting with a freebody diagram of point A) gives the following results:

| MEMBER | LENGTH (ft) | AREA (in²) | E (psi) | p (lb) | P (lb) |
|---|---|---|---|---|---|
| BEAM | 24. | 25.375 | $1.6 \times 10^6$ | $-2.0$ | $-2.0\, P_{king\ post}$ |
| ROD 1 | 12.37 | 0.785 | $30 \times 10^6$ | 2.06 | $2.06\, P_{king\ post}$ |
| ROD 2 | 12.37 | 0.785 | $30 \times 10^6$ | 2.06 | $2.06\, P_{king\ post}$ |
| KING POST | 3. | 12.25 | $1.6 \times 10^6$ | $-1.0$ | $-1.0\, P_{king\ post}$ |

The deflection due to a unit load in the king post is equal to the sum of the PpL/AE terms.

BEAM: $\dfrac{(-2\, P_{kp})(-2)\, lb^2\, (24)\, ft\, (12)\, in/ft}{(25.375)\, in^2\, (1.6 \times 10^6)\, lb/in^2} = 2.837 \times 10^{-5}\, (P_{king\ post})$

RODS: $\dfrac{(2)(2.06\, P_{kp})(2.06)(12.37)(12)}{(.785)(30 \times 10^6)} = 5.35 \times 10^{-5}\, (P_{king\ post})$

KING POST: $\dfrac{(-1)(-1 P_{kp})(3)(12)}{(12.25)(1.6 \times 10^6)} = 1.837 \times 10^{-6}\, (P_{king\ post})$

$$\text{TOTAL} = 8.371 \times 10^{-5}\, (P_{king\ post})$$

The compatibility equation is, therefore,

$(7.56)\, in - (2.80 \times 10^{-3})(P_{king\ post}) = (8.371 \times 10^{-5})(P_{king\ post})$

Solving for the force in the king post,

$P_{king\ post} = (7.56)/(8.371 \times 10^{-5} + 2.80 \times 10^{-3}) = 2{,}622\, lb$

As with the previous problem, a quick check on the upper and lower bounds for this indeterminate solution is worth the effort. If the post were rigid, the force would be 5/8 of the applied load, or 2700 pounds. If the beam were very flexible, the force would be 1/2 the applied load, or 2160 pounds. Since the solution to the compatibility equation falls between these bounds, it seems reasonable.

The deflection at the center is

$$(3.293 \times 10^{-4})(2,622) = 0.86 \text{ inches}$$

The deflection-to-span length ratio is $(24)(12)/(0.86) = 334$, an acceptable amount of deflection.

In order to evaluate the stresses in the beam, first solve for the reactions. Then find the maximum bending and shear stresses. The beam is loaded in axial compression, as well as bending, so the two longitudinal stresses must be superimposed.

End reactions = $[(180) \text{ lb/ft } (24) \text{ ft} - (2,622) \text{lb}]/(2) = 849 \text{ lb}$

Moment at the king post is

$$[(849)\text{lb}(12)\text{ft} - \{(180)\text{lb/ft}(12)^2 \text{ ft}^2/(2)\}](12)\text{in/ft} = 33,264 \text{ in-lb}$$

Maximum positive moment occurs at the point where the shear is zero.

$V = 0$ @ $(849) \text{ lb}/(180) \text{ lb/ft} = 4.72 \text{ ft}$

$$M_{4.72'} = [(849)(4.72) - \{(180)(4.72)^2/(2)\}](12) = 24,027 \text{ in-lb} = M_{max}$$

The section modulus, 30.661 in$^3$, is found in appendix 5.

The maximum bending stress is

$$M_{max}/S = (24,027)/(30.661) = 784 \text{ psi}$$

The axial forces have to be determined in order to find the axial stresses in the rods and beam. Look first at the equilibrium of the bottom of the king post.

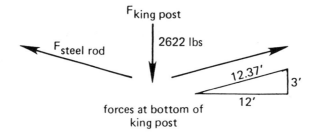

forces at bottom of
king post

The force in a rod is

$$P_{rod} = \frac{2,416 \text{ lb}}{2} \times \frac{12.37 \text{ ft}}{3 \text{ ft}} = 4,981 \text{ lb}$$

The king post force was divided in half because there are two rods, and hence, two vertical rod components.

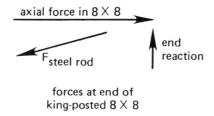

forces at end of
king-posted 8 × 8

The axial force in the beam is

$$P_{beam} = \frac{(4981) \text{lb} (12) \text{ft}}{(12.37) \text{ft}} = 4832 \text{ lb}$$

The axial compressive stress in the beam caused by this axial load is

$$f_a = (4832) \text{ lb}/(25.375) \text{ in}^2 = 190 \text{ psi}$$

The maximum combined compressive stress, therefore, occurs at the top of the beam, 4.72 feet from either end, and is equal to,

(190) psi + (784) psi = 974 psi

This is less than the allowable bending stress of 1400 psi. Using the interaction equation for combined bending and axial compression would be justified in this problem. However, since only the allowable bending stress was given, a simple superposition was the only solution method possible.

The axial stress in the steel rods is

$$f_a = (4981) \text{ lb}/ (0.785) \text{ in}^2 = 6,340 \text{ psi} < 20,000 \text{ psi allowable}$$

The king posted beam meets allowable stress requirements in the steel rods and the 4x8 beam. The maximum deflection is less than the span divided by 334, meeting most deflection criteria.

## SAMPLE PROBLEM 3

Investigate the bending and shear stresses in the glued-laminated timber shown. Use the National Design Specifications of the National Forest Products Association. Specify the edition used in your solution. Neglect the self-weight of the beam.

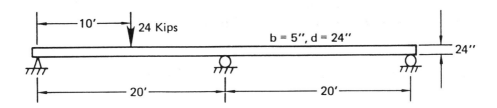

Given: Douglas fir, 24FV5, normal duration of load, dry conditions b = 5 inches, d = 24 inches (actual dimensions)
Beam supported laterally only at the supports.
For the purposes of this problem use $L_u$ as that of a simple span.

Using handbook solutions, evaluate the maximum bending and shear stresses in the beam. Compare these with the allowable stresses. The allowable bending stress will be determined either by lateral buckling or extreme fiber stress.

The maximum shear occurs between the load and the central support, and is equal to (19/32) P (as determined from beam tables for 2-span continuous beams).

$$V_{max} = (19/32)P = (19/32)(24,000) \text{ lb} = 14,250 \text{ lb}$$

The maximum shear stress is, therefore,

$$f_v = (14,250) \text{ lb } (1.5)/(5)(24) \text{ in}^2 = 178 \text{ psi}$$

Appendix 2 gives an allowable shear stress, $F_v$, of only 155 psi for V5 lay-up. The beam is overstressed in shear.

The maximum moment occurs under the load, and is equal to (13/64) PL, again as determined from beam tables for 2-span continuous beams.

$$M_{max} = (13/64)PL = (13/64)(24,000) \text{ lb } (20) \text{ ft } (12) \text{ in/ft}$$
$$= 1,170,000 \text{ in-lb}$$

Since the given dimensions are actual sizes, the section modulus is calculated as

$$S = bd^2/6 = (5)(24)^2/(6) = 480 \text{ in}^3,$$

$$f_b = M/S = (1,170,000)/(480) = 2,440 \text{ psi}.$$

In order to find the allowable bending stress, both lateral buckling and extreme fiber stresses must be considered.

Lateral Buckling:
$$L_u/d = (20) \text{ ft } (12) \text{ in/ft}/(24) \text{ in} = 10.0$$

$L_u/d$ is not 17, so it will be necessary to correct the calculated effective length.

The effective length, $L_e$, is equal to

$$L_e = (1.61)(20) \text{ ft } (12) \text{ in/ft } [(0.85) + (2.55)/(10.0)] = 427 \text{ in}$$

The 1.61 factor comes from figure 40, and the correction factor in square parentheses is explained beneath figure 40. The slenderness ratio, $C_s$, is calculated from equation 23.

$$C_s = [(427) \text{ in } (24) \text{ in}/(5)^2 \text{ in}^2]^{0.5} = 20.25$$

This is larger than 10, so the beam is not short. In order to evaluate $C_k$, E and $F_b$ from appendix 2 are needed.

$$F_b = 2,400 \text{ psi}, \quad E = 1,700,000 \text{ psi}$$

In the 1982 and subsequent editions of the National Design Specifications, there are many more glulam layup combinations to choose from than in any of the earlier editions. The value, $C_k$, which separates long and intermediate beams is, therefore,

$$C_k = (0.811)[(1,700,000) \text{ psi}/(2,400) \text{ psi}]^{0.5} = 21.58$$

This beam is barely in the intermediate range. The allowable bending stress, considering lateral buckling, is

$$F_b' = (2400)[1 - (1/3)(20.25/21.58)^4] = 1,780 \text{ psi}$$

This beam is significantly overstressed in bending, but check the extreme fiber stress for completeness. The only factor which applies to this case is the size factor, $C_F$. Table 18 gives this as 1.01, so the beam is just about at the allowable stress at the extreme fibers, but it is significantly overstressed when lateral stability is considered.

## SAMPLE PROBLEM 4

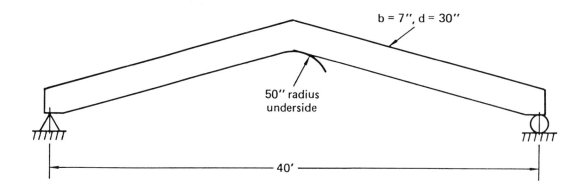

Investigate only the bending and radial stresses in the glued-laminated timber beam shown, according to the 1986 National Design Specifications of the National Forest Association.

Given:   Total dead plus live load of 1000 lbs per foot.
         Normal load duration, dry use.
         Douglas fir 24 F, b = 7 inches, d = 30 inches
         Laminations one half inch thick
         Beam supported laterally at ends and mid span.

The actual bending and radial stresses must be compared with allowable stresses that are not easily determined. The allowable bending stress is either a function of lateral stability or of extreme fiber stresses. Both cases will be investigated, with the lower value controlling. The allowable radial stress depends on whether the stress is tension or compression, and is found in the text accompanying table 19.

Bending Stresses:

The maximum moment occurs at the centerline and is equal to $wL^2/8$.

$M_{max}$ = (1000) lb/ft (40)² ft² (12) in/ft/(8) = 2,400,000 in-lb

S = $bd^2/6$ = (7)(30)²/(6) = 1050 in³

$f_b$ = M/S = (2,400,000)/(1050) = 2,286 psi

# Timber Design

Now, evaluate the two different allowable bending stresses. First, consider lateral stability.

$$L_u = (40) \text{ft}/(2) = 20 \text{ ft}$$

$$L_u/d = (20)(12)/(30) = 8.0 \ (\neq 17)$$

Figure 40, case 2 gives an effective unbraced length of

$$L_e = (1.92)(20) \text{ ft } [(0.85) + (2.55)/(8.0)] = 44.88 \text{ ft}$$

The slenderness factor, $C_s$, is

$$C_s = [(44.88) \text{ ft } (12) \text{ in/ft } (30) \text{ in}/(7)^2 \text{ in}^2]^{0.5} = 18.16 \ (>10)$$

In order to find $C_k$, the dividing line between long and intermediate beams, $F_b$ and E are required. Appendix 2 gives several possible values for "24 F" Douglas fir. The 1/2" laminae imply that this is a fairly high quality member, so it is not out of line to assume the stiffest E of 1,800,000 psi. This yields a $C_k$ of

$$C_k = (0.811)[(1,800,000)/(2,400)]^{0.5} = 22.21$$

Since $10 < C_s < C_k$, this is an intermediate beam and $F_b'$ is

$$F_b' = (2400)[1 - (1/3)(18.16/22.21)^4] = 2042 \text{ psi}$$

The beam is overstressed considering lateral stability, but check the extreme fiber allowable stress, for completeness.

The size factor, $C_F$, from table 18, is 0.90. The curvature factor, $C_c$, for the region of curved laminae--which is also the region of highest bending stress--is calculated with equation 27.

$$C_c = 1 - (2000)(.5/50)^2 = 0.80$$

The use is dry, so the CUF is 1.0, resulting in an allowable extreme fiber bending stress of

$$F_b = (2400) \text{ psi } (0.90)(0.80)(1.0) = 1728 \text{ psi}$$

This is actually smaller than the allowable stress controlled by lateral stability. This beam is significantly overstressed according to the 1986 National Design Specifications.

The radial stress is evaluated with equation 28, and is equal to

$$f_r = 3M/2Rbd = (3)(2,400,000)\text{in-lb}/(2)(65)\text{in}(30)\text{in}(7)\text{in}$$
$$= 264 \text{ psi (tension)}$$

The applied bending moment decreases the curvature (increases the radius of curvature), straightening the member out at the bend. This makes the radial stress a tensile one. In Douglas fir the allowable radial stress in tension is only 15 psi, so the beam is highly overstressed in radial tension at the midspan bend. If it is essential to use this species and configuration, mechanical radial reinforcement should be added. This reinforcement could either be through-bolts or lag screws, sized to have sufficient capacity to handle all the radial forces.

## SAMPLE PROBLEM 5

Design a timber beam to carry two concentrated loads of 800 pounds each, spaced 5 feet from each other, acting anywhere on a simply supported span of 20 feet. Allowable extreme fiber stress parallel to grain is 1500 psi, and allowable horizontal shear is 40 psi. Only 2x6, 2x8, 2x10, 2x12, 2x14 nominal size boards are available. Any size beam may be built up using the available boards. Neglect the dead weight of the beam.

The basic design procedure is to evaluate the maximum bending moment and shear by moving the applied load. Dividing by the allowable stresses will give the required area and section modulus. Various combinations of boards will be investigated, with a minimum area solution being chosen as the most economical.

The practicality of combining different size boards in one beam is somewhat suspect, and should only be done with care. If the number of boards turns out to be large, then consideration should be given to assembling them in more efficient configurations than "stacked." An I beam or box beam, for instance, might be worth the effort in extreme situations.

Since loads within a beam depth of a support are neglected when calculating shears, the maximum shear will occur when one of the loads is one beam depth from an end. The beam depth is an unknown at this point, but an assumption of 1 foot should not be too far from the eventual design.

The maximum shear, therefore, is

$$V_{max} = [(800) \text{ lb } (19) \text{ ft} + (800) \text{ lb } (14) \text{ ft}]/(20) \text{ ft} = 1,320 \text{ lb}$$

Most structural handbooks include maximum moment formulas for two moving loads. The formula used depends on the load spacing relative to the beam length. The load spacing, a, is 5 feet. This is compared with a load spacing which separates beam behaviors.

$$(20) \text{ ft } [(2) - (2)^{.5}] = 11.72 \text{ feet}$$

Since the actual spacing is less than the dividing value, the following formula applies:

$$\begin{aligned} M_{max} &= (P/2L)(L - a/2)^2 \\ &= [(800) \text{ lb}/(2)(20) \text{ ft}] [(20) \text{ ft} - \{(5) \text{ ft}/(2)\}]^2 (12) \text{ in/ft} \\ &= 73,500 \text{ in-lb} \end{aligned}$$

To calculate the required area and section modulus, modify the allowable stresses first for actual conditions. Assume that the use conditions are dry and load duration is normal, since the problem statement mentions nothing to the contrary. The National Design Specifications allow a 50% increase in allowable shear stress if the maximum shear force is evaluated in a systematic way. Given the very low allowable shear stress in this problem and the care spent in finding the maximum shear, this 50% increase is justifiable. The required section properties, therefore, are

$$A_{req'd} \geq (1.5)(1,320) \text{ lb}/(1.5)(40) \text{ psi} = 33.0 \text{ in}^2 = bd$$

The first 1.5 comes from equation 29, the second is the 50% increase in allowable shear stress.

$$S_{req'd} \geq (73,500) \text{ in-lb}/(1500) \text{ psi} = 49.0 \text{ in}^3 = bd^2/6$$

Since a simple built-up beam will have a width which is a multiple of 1.5 inches, solve for the required depths for bending and shear at a range of potential widths.

| b | 3" | 4.5" | 6" | 7.5" |
|---|---|---|---|---|
| $d_V$ | 11.0" | 7.33" | 5.5" | 4.4" |
| $d_M$ | 9.9" | 8.08" | 7.0" | 6.26" |

Avoiding 2x14's as being too rare and expensive, the following combinations will provide the depths required:

| 2-2x12's | Area = (2)(16.875) = 33.75 in$^2$ |
| 3-2x10's | Area = (3)(13.875) = 41.63 in$^2$* |
| 4-2x8's  | Area = (4)(10.875) = 43.50 in$^2$* |

Since lumber cost is largely a function of material volume, minimizing the cross-sectional area is the selection criterion. Therefore, use two 2x12's.

The problem statement does not include any mention of deflections, so they will be neglected. The beam designed is small enough that more complicated, but efficient, configurations are not justifiable.

\* Could qualify as repetitive use.

# SAMPLE PROBLEM 6

A 40 foot simple span, 6.75"x30", 24F-V5 Glulam beam (western species wood used in fabrication), braced laterally on 10 foot centers, normal load duration, dry use conditions, carries a dead load of 100 plf and a live load of 700 plf.

Determine:
1) Section properties (include any size or form factors)
2) Relevant tabulated allowable stresses
3) Design moment, shear, and reactions
4) Deflections under dead and total loads, with suggested camber
5) Allowable moment, shear, and minimum support length

1) The member dimensions are actual. The section properties are, therefore,

$A = (6.75)(30) = 202.5 \text{ in}^2$
$S = (6.75)(30)^2/(6) = 1,012.5 \text{ in}^3$
$I = (6.75)(30)^3/(12) = 15,187.5 \text{ in}^4$
$C_F = (12/d)^{1/9} = (12/30)^{1/9} = 0.903$

2) The member is loaded principally in bending about the X-X axis, so appendix 2 gives the following allowable stresses:

$F_b$ = 2400 psi (assume beam installed right-side-up)
$F_{c\perp}$ = 650 psi (support load is on lower, tension face)
$F_v$ = 155 psi
$E$ = 1,700,000 psi

3) $M = wL^2/8 = (800) \text{ plf } (40)^2 \text{ ft}^2 (12) \text{ in/ft}/(8)$
$\qquad = 1,920,000 \text{ in-lb}$

$V = (800) \text{ plf } [\{(40) \text{ ft}/(2)\} - \{(30) \text{ in }/(12) \text{ in/ft}\}] = 14,000 \text{ lb}$

Note that the load within a beam depth of a support was neglected when calculating the maximum shear.

$R = (800) \text{ plf } (40) \text{ ft}/(2) = 16,000 \text{ lb}$

4) $D_{max} = 5wL^4/384EI$

$$D_{dead} = \frac{5(100)\text{plf }(40)^4\text{ft}^4\text{ }(1728)\text{in}^3/\text{ft}^3}{384(15,187.5)\text{in}^4\text{ }(1,700,000)\text{psi}} = 0.22 \text{ in}$$

$D_{live} = (.22) \text{ in } (800/100) = 1.76 \text{ in}$

Suggested camber is 1.5 the dead load camber or, $(1.5)(.22) = .33$ in, say 3/8" at the centerline.

5) The allowable bending stress will be governed by stability or fiber stress. First check stability.

$L_e = (1.92)(10) \text{ ft } (12) \text{ in/ft} = 230.4 \text{ in, (case 2, figure 40)}$

$C_s = [(230.4) \text{ in } (30) \text{ in}/(6.75)^2 \text{ in}^2]^{0.5} = 12.32$

Since $C_s$ is > 10, the beam is not short.

$C_K = (.811)[(1,700,000)/(2,400)]^{0.5} = 21.58$

$C_s$ is less than $C_K$ and greater than 10, so the beam is an intermediate one.

$F_b' = (2400)[1 - (1/3)(12.32/21.58)^4] = 2315. \text{ psi}$

Now check fiber stress.

$F_b = (0.903)(2400) = 2167 \text{ psi}$

Stresses in the extreme fibers controls allowable bending stresses.

$M_{all} = F_bS = (2400) \text{ psi } (.903) (1,012.5) \text{ in}^3 = 2,190,000 \text{ in-lb}$

$V_{all} = F_vA/1.5 = (155) \text{ psi } (202.5) \text{ in}^2/(1.5) = 20,925 \text{ lb} > 14,000 \text{ lb}$

The beam is adequate in bending and shear stress considerations.

The required bearing length may be determined by trial and error or by solving explicitly for the minimum required length. This is done by equating the allowable and actual bearing stresses, both as functions of the bearing length, $l_b$.

$$(650)(l_b + .375)/l_b = (16,000)/(6.75)l_b$$

The solution to this equation is $l_b = 3.27"$

Use a 3.5" bearing length.

## SAMPLE PROBLEM 7

Determine two acceptable S4S sections for a sheltered 19 foot long column. The imposed design loads are a 4,000 pound axial snow load with a 3 inch eccentricity, and a 150 pound/foot uniform, transverse wind load. Assume a lateral support is provided at midspan, in a direction perpendicular to the transverse load only. The species and grade available provides these design properties:

$F_c = 1,300$ psi
$F_b = 1,600$ psi (single use)
$E = 1,800,000$ psi

This beam-column is subjected to eccentric axial and transverse loads. Equation 15 is the interaction formula used to evaluate combined stress levels in this general case.

Both actual and allowable stresses will vary with member size. The design method, therefore, is a trial-and-error iteration with a minimum-size member as the goal. With snow and wind loading there are two load duration factors to consider. Design the column for the combined loading and check its capacity against the individual cases.

Making a good first size estimate can save time. Given the lateral bracing, look for sections roughly twice as deep as they are wide.

The maximum bending moment is

$$[(150)\text{plf } (19)^2 \text{ ft}^2/(8)](12) \text{ in/ft} + (4,000)\text{lb } (3)\text{in} = 93,225 \text{ in-lb}$$

Note that the eccentricity has been conservatively assumed to create bending about the same axis as does the transverse loading.

Try a 6x10; 5.5"x9.5", A = 52.25 in$^2$, S = 82.729 in$^3$ (appendix 5)

First, evaluate the actual stresses.

$$f_c = (4,000) \text{ lb}/(52.25) \text{ in}^2 = 76.6 \text{ psi}$$

$$f_b = (93,225) \text{ in-lb}/(82.729) \text{ in}^3 = 1,126.9 \text{ psi}$$

It is assumed that the lateral support will be used to brace the weak axis of buckling.

Next, establish the allowable stresses.

$F_b$: First consider lateral buckling.

$$L_e = (19/2)\text{ft } (12)\text{in/ft } (1.92) = 218.9 \text{ in}$$

The 1.92 factor is the effective length of the beam, from figure 40.

The slenderness factor, $C_s$ is

$$= [(218.9)\text{in } (9.5)\text{in}/(5.5)^2 \text{in}^2]^{.5} = 8.28 < 10$$

Since $C_s$ is less than 10, the beam-column acts as a short beam. This is often true of the beam action of a beam-column.

$$F_b' = (1.33)(1,600) \text{ psi} = 2,128 \text{ psi}$$

The 1.33 wind load duration factor comes from table 2. Since the member is less than 12 inches deep, there is no size effect to consider.

$F_c$: Determine which axis controls in buckling.

strong axis: $L_e/d = (1.0)(19)\text{ft }(12)\text{in/ft }/(9.5)\text{in} = 24.00$

weak axis: $L_e/d = (1.0)(19/2)\text{ft }(12)\text{in/ft }/(5.5)\text{in} = 20.74$

## Timber Design

The 1.0 effective length factor is found in table 16. Strong axis buckling controls in this member size. $C_s$ is 24.00 in this case.

$$K = (0.671)[(1,800,000)psi /(1,300)psi (1.33)]^{.5} = 21.65$$

Note that the 1.33 load duration factor appears in this equation. The K factor is a material property, and it will not be recalculated for any subsequent member sizes.

Since $C_s$ is greater than K, the beam-column acts as a long column.

$$F_c' = (0.3)(1,800,000)psi/(24.00)^2 = 937.5 \text{ psi}$$

$$J = 1.0$$

Substitute these values into the interaction equation (equation 15).

$$\frac{(76.6)psi}{(937.5)psi} \pm \frac{(1,126.9)psi + (76.6)psi[6 + (1.5)(1.0)](3)in/(9.5)in}{(2,128)psi - (1.0)(76.6) psi}$$

$$= 0.08 + 0.62 = 0.70 \quad 1.0$$

The 6x10 is adequate. As a check, the reader should perform a similar calculation for a 6x8. The interaction formula shows that that member would be 20% overloaded.

The second section should be a 4 inch wide member. A 4x14 has a smaller area than the 6x10, but will be difficult to find in most locations.

Try a 4x12; 3.5"x11.25", A = 39.375 in$^2$, S = 73.828 in$^3$

Evaluate the actual stresses.

$$f_c = (4,000)lb/(39.375) in^2 = 101.6 \text{ psi}$$

$$f_b = (93,225)in\text{-}lb /(73.828) in^3 = 1262.7 \text{ psi}$$

Establish the allowable stresses.

$F_b$: First consider lateral buckling.

The slenderness factor, $C_s$ is

$$= [(218.9)\text{in }(11.25)\text{in}/(3.5)^2\text{in}^2]^{.5} = 14.18 < 10$$

Since $C_s$ is greater than 10, the beam-column is not a short beam.

$$C_k = (0.811)[1,800,00/(1.33)(1,600)]^{.5} = 23.59$$

The beam-column acts as an intermediate beam.

$$F_b' = (1.33)(1,600)\text{psi }[1 - 1/3(14.18/23.59)^4]$$
$$= 2,035.4 \text{ psi}$$

The 1.33 wind load duration factor comes from table 2. Since the member is less than 12 inches deep, there is no size effect to consider.

$F_c$: Determine which axis controls in buckling.

  strong axis: $L_e/d = (1.0)(19)\text{ft }(12)\text{in/ft}/(11.25)\text{in} = 20.27$
  weak axis: $L_e/d = (1.0)(19/2)\text{ft }(12)\text{in/ft}/(3.5)\text{in} = 32.57$

Weak axis buckling controls in this member size. $C_S$ is 32.57 in this case. Since $C_S$ is greater than K, the beam-column acts as a long column.

$$F_c' = (0.3)(1,800,000)\text{psi}/(32.57)^2 = 509.0 \text{ psi}$$
$$J = 1.0$$

Substitute these values into the interaction equation.

$$\frac{(101.6)\text{psi}}{(509.0)\text{psi}} + \frac{(1,262.7)\text{psi} + (101.6)\text{psi }[6 + (1.5)(1.0)](3)\text{in}/(11.25)\text{in}}{(2,035.4)\text{psi} - (1.0)(101.6)\text{psi}}$$

$$= 0.20 + 0.76 = 0.96 < 1.0$$

The 4x12 is adequate, but a 4x8 would clearly be overloaded and is not worth checking.

Now consider for the loads acting individually. Wind load alone would not control because the LDF would not change, but the moment and axial stresses would be decreased. Nor will snow load alone control, because both sections behave as long columns, on which the LDF has no influence. Therefore, $F_c'$ will be unchanged and only the moment will be decreased because the wind load is omitted.

Either a 4x12 or a 6x10 will satisfy the design requirements.

## SAMPLE PROBLEM 8

What is the wind load capacity of the 1/2" diameter bolts in the brace shown?

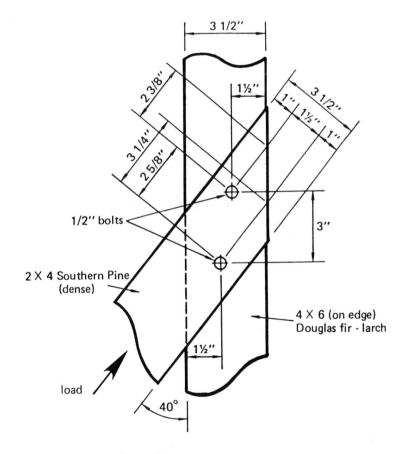

The single shear connection has members which are not aligned with each other, of different thickness and species. Evaluate the capacity of the bolts in each member as though both members were of the same species. The lowest value will control. Then, adjust for number of connectors, load duration, and end/edge spacings to establish the wind load capacity.

First, check capacities as though both members were dense Southern pine.

In the 4x6, the capacity is

$$\frac{(.5)(1490)\text{lb }(1010)\text{lb}}{(1490)\text{lb}[\sin^2(40°)] + (1010)\text{lb}[\cos^2(40°)]} = 623 \text{ lb}$$

The .5 accounts for single shear. Hankinson's formula was used with the values of table 13 because the load is applied at 40° to the grain of the 4x6. The bolt actually has 5.5" in the main member, but table 13 has no value for a 1/2" bolt that long. The longest length included for a 1/2" bolt is 4", which yields the values used above.

In the 2x4, the capacity is

(.5)(1490)lb = 745 lb

This value is for a length of (2)(1.5)in = 3 in. The doubled thickness is used because this is the thinner member of the two. Now check the capacities as though both members were Douglas fir-larch.

In the 4x6, the capacity is

$$\frac{(.5)(1270)\text{lb }(1010)\text{lb}}{(1270)\text{lb}[\sin^2(40°)] + (1010)\text{lb}[\cos^2(40°)]} = 574 \text{ lb}$$

In the 2x4, the capacity is

(.5)(1270) lb = 763 lb

So far, the 574 pound capacity of the bolt in the 4x6 appears to control.

Check reductions for number of members. Table 8.A shows that there is no reduction in capacity for two bolts, no matter how they are arranged. The wind load duration factor of 1.33, from table 2, will apply.

Now, check the spacings for any possible reduction of the connection capacity.

First check spacings in the 4x6. Establish whether the two bolts are considered as one two-bolt row or two one-bolt rows.

$$d = 1.5" < 3/4" = 3"/4 = s/4$$

Therefore, the bolts are considered a single vertical row of two bolts, in the 4x6. The bolt spacing is 3", which is equal to 6 bolt diameters -- 2 more than the required 4 diameters.

The end distance is meaningless in the long 4x6. The edge distance is complicated by the 40° load angle. If the load were parallel with the grain, 1-1/2 diameters would be the required edge distance. If the load were perpendicular to the grain, 4 diameters would be required. One way to deal with the load angle between parallel and perpendicular is to apply the Hankinson formula to the edge distance.

$$\text{Edge Distance} = \frac{(4)\text{diameters}\ (1.5)\text{diameters}}{(4)\sin^2(40°) + (1.5)\cos^2(40°)} = 2.37 \text{ diameters}$$

The required edge distance is $(2.37)(.5)\text{in} = 1.18$ inches, less than the 1.5 inches provided. Note that this edge distance is the loaded edge distance. It should be provided at both edges of the 4x6 with a wind load which is expected to apply in both directions.

Now, look at the spacings in the 2x4. First check whether the two bolts are two one-bolt rows, or one two-bolt row.

$$d = 1.5" > 0.81" = 3.25"/4 = s/4$$

The two bolts can be viewed as two rows of one bolt each. The required spacing between rows of bolts, with load parallel with the grain, is 3 diameters, or 1.5 inches. The spacing provided is 1.5 inches, which is adequate.

With load parallel with the grain, the edge distance should be 1.5 diameters, or .75 inches. The 1 inch provided is adequate.

The end distance required for full design capacity in tension is 7 diameters, or 3.5 inches. Neither bolt has this required end distance, but a linear reduction in capacity is allowed for less than full end distance -- so long as at least one-half the required is provided. Calculate the reduced capacity of the bolts, considering their individual end distances.

$$(574)lb\ [(3.25)in/(3.5in + (2.375)in/(3.5)] = 923\ pounds$$

A case could be made for using the 745 pound capacity of the Southern pine 2x4 as the basic capacity in the above equation. It is conservative to use the 574 pounds, and such usage better accounts for the complex interaction between bolted members.

Finally, use the 1.33 load duration factor of table 2 to find the wind load capacity for this connection.

$$(1.33)\ (922)lb = 1,230\ pounds$$

# SAMPLE PROBLEM 9

An assembled lumber box beam is shown. What is the shear capacity of the section, considering load in the nails and stress in the lumber?

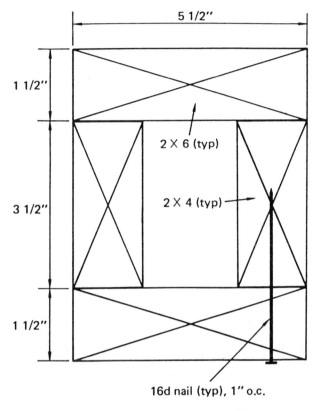

Douglas fir - larch #1

A distinction will be drawn between shear stress and shear flow in the analysis of this section. Shear flow is the shear that crosses a plane in the section, irrespective of the width of that plane. The maximum stress in the lumber is at the neutral axis, which means that two Q's will be calculated.

Both shear calculations require the section moment of inertia.

$$I = (2)(1.547)in^4 + (2)(8.25)in^2[(3.5+1.5)in/(2)]^2 + (2)(5.359)in^4$$
$$= 116.94 \ in^4$$

Appendix 5 gives the areas and moments of inertia used above. Note that the 2x6 used flat is the same as a 6x2 in appendix 5. The parallel axis theorem was used to account for the displacement of the 2x6 flanges from the neutral axis.

The shear flow, q, is found with a form of the standard shear stress equation.

$$q = VQ/I$$

Rearranging, $V_{all} = \dfrac{q_{all} I}{Q}$

The allowable shear flow is the product of the nail spacing and capacity.

$$q_{all} = (1)\text{nail/inch/side }(2)\text{sides }(108)\text{lb/nail} = 216 \text{ lb/in}$$

Table 3 shows Douglas fir to be a group 2 species. Table 6 gives the capacity for the group 2 species. Note that both lines of nails are included in the connection capacity.

Next, calculate the Q at the nail line.

$$Q_{nail\ line} = (8.25)in^2[(3.5)in + (1.5)in]/(2) = 20.63 \text{ in}^3$$

This quantity is simply the moment of the area of the 2x6 about the neutral axis.

Calculate the shear capacity, as limited by the nails.

$$V_{all} = (216)\text{lb/in}(116.94)in^4/(20.63)in^3 = 1,224 \text{ lb}$$

The lumber shear stress evaluation takes into account the width of the plane at which the stresses are being calculated. Rearranging the standard equation gives the allowable shear capacity.

$$V_{all} = F_v It/Q_{max}$$

The maximum Q is found at the neutral axis. It will be equal to the Q found earlier plus the two upper halves of the webs.

$$Q_{max} = (20.63)in^3 + (2)[(1.5)in(3.5)in/(2)][(3.5)in/(4)]$$
$$= 25.22 \text{ in}^3$$

Appendix 2 gives a $F_v$ of 95 psi for Douglas fir-larch.

$$V_{all} = [(95)psi\ (116.94)in^4\ (2)\ (1.5)in]\ /(25.22)in^3 = 1,322\ lb$$

The nail capacity controls and limits the shear capacity of the assembled section to 1,224 pounds. The nail spacing could be increased in regions of the beam where the shear is reduced.

## SAMPLE PROBLEM 10

Find the maximum allowable combined wind and dead loads on the 2x12 rafter shown. Determine the maximum notch depth allowed at the ends of the rafter. The deflection is to be limited to L/180. Use kiln-dried #2 Southern pine.

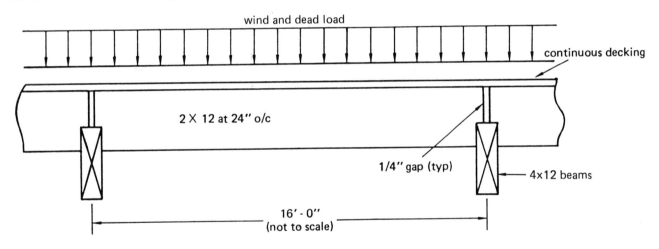

With a rafter this long, bending will most likely control the load capacity. Determine the capacity with bending stresses and check shear, bearing, and deflection.

Appendix 2 gives the following material properties:
$F_b$ = 1,500 psi (repetitive use)
$F_v$ = 95 psi
$E$ = 1,600,000 psi
$F_{c\perp}$ = 565 psi

Table 2 yields a load duration factor of 1.15, which applies only to the bending and shear stresses -- not to E or $F_{c\perp}$.

Appendix 5 gives the section properties for the 2x12.

The bending capacity is limited by the maximum moment.

$$w = \frac{F_b 8S}{L^2} = \frac{(1.15)(1,500)\text{psi}(8)(31.641)\text{in}^3(1)\text{ft}/(12)\text{in}}{(16)^2 \text{ft}^2}$$

$$= 142.1 \text{ lb/ft}$$

Check the bearing stress with an end reaction for this maximum load.

$$R = (142.1)\text{lb/ft}(16)\text{ft}/(2) = 1,137 \text{ lb}$$

Determine the bearing length provided.

$$L_b = [(3.5)\text{in} - (.25)\text{in}]/(2) = 1.625 \text{ in}$$

$$f_{c\perp} = (1,315)\text{lb} / (1.625)\text{in}(1.5)\text{in} = 540 \text{ psi} < 565 \text{ psi}$$

Determine the minimum required effective depth required at the ends of the rafter. Solve equation 30 for d'.

$$d' = \{3Vd/2bF_v\}^{.5}$$
$$= \{(3)(1,137)\text{lb}(11.25)\text{in}/(2)(1.5)\text{in}(95)\text{psi}(1.15)\}^{.5}$$
$$= 10.82 \text{ in}$$

This minimum required depth means the maximum notch depth is 3/8".

Check the deflection.

$$Y_{max} = 5wL^4/384EI$$
$$= \frac{(5)(142.1)\text{lb/ft}(16)^4\text{ft}^4(1,728)\text{in}^3/\text{ft}^3}{(384)(1,600,000)\text{psi}(177.979)\text{in}^4}$$
$$= 0.74 \text{ in}$$

This deflection is converted into a fraction of the span.

(16)ft (12)in/ft /(.74)in = 256

The deflection is equal to L/259 < L/180.

Bending limits the snow and dead load to 142.1 lb/ft. This load per length of beam is converted to an equivalent distributed load.

(142.1)lb/ft / (2)ft on center = 71 psf

There are at least two further refinements possible in this problem. The span could be decreased from the assumed center-to-center of the glulam beams. The span could be justified as the face-to-face clear distance, plus the distance to the centers of bearing area required to handle the actual reactions. The shear could be further reduced by neglecting the distributed load within a beam depth of the glulam members. Neither of these refinements would have much effect. Given the unpredictable response of notched beams, conservatively neglecting these effects is reasonable.

**RESERVED FOR FUTURE USE**

## 12. APPENDICES

## Appendix 1

### END GRAIN IN BEARING

Design values for end-grain bearing parallel to grain on a rigid surface $F_g$ in pounds per square inch

| Species | Wet service conditions[1] | Dry service conditions[1] Sawn lumber[2] More than 4" thick | Dry service conditions[1] Sawn lumber[2] Not more than 4" thick[3] | Glued laminated timber |
|---|---|---|---|---|
| Ash, Commercial White | 1370 | 1510 | 2060 | 2400 |
| Aspen | 740 | 820 | 1110 | 1300 |
| Balsam Fir | 1140 | 1250 | 1710 | 1990 |
| Beech | 1190 | 1310 | 1780 | 2080 |
| Birch, Sweet and Yellow | 1150 | 1260 | 1720 | 2010 |
| Black Cottonwood | 620 | 690 | 930 | 1090 |
| California Redwood | 1560 | 1720 | 2270 | 2620 |
| California Redwood, Open grain | 1150 | 1270 | 1670 | 1940 |
| Coast Sitka Spruce | 950 | 1040 | 1420 | 1660 |
| Coast Species | 950 | 1040 | 1420 | 1660 |
| Cottonwood, Eastern | 760 | 840 | 1150 | 1340 |
| Douglas Fir - Larch (Dense)[4] | 1570 | 1730 | 2360 | 2750 |
| Douglas Fir - Larch[4] | 1340 | 1480 | 2020 | 2350 |
| Douglas Fir South | 1220 | 1340 | 1820 | 2130 |
| Eastern Hemlock - Tamarack[4] | 1150 | 1270 | 1730 | 2020 |
| Eastern Hemlock | 1140 | 1260 | 1710 | 2000 |
| Eastern Softwoods | 890 | 980 | 1340 | 1560 |
| Eastern Spruce | 890 | 980 | 1340 | 1560 |
| Eastern White Pine[4] | 900 | 1000 | 1360 | 1580 |
| Eastern Woods | 740 | 820 | 1110 | 1300 |
| Engelmann Spruce - Alpine Fir | 810 | 890 | 1220 | 1420 |
| Hem - Fir[4] | 1110 | 1220 | 1670 | 1940 |
| Hickory and Pecan | 1370 | 1510 | 2050 | 2400 |
| Idaho White Pine | 930 | 1020 | 1390 | 1630 |
| Lodgepole Pine | 970 | 1060 | 1450 | 1690 |
| Maple, Black and Sugar | 1140 | 1260 | 1710 | 2000 |
| Mountain Hemlock | 1070 | 1170 | 1600 | 1870 |
| Mountain Hemlock - Hem Fir | 1070 | 1170 | 1600 | 1870 |
| Northern Aspen | 740 | 810 | 1110 | 1290 |
| Northern Pine | 1040 | 1150 | 1570 | 1830 |
| Northern Species | 880 | 970 | 1320 | 1540 |
| Northern White Cedar | 740 | 810 | 1110 | 1290 |
| Oak, Red and White | 1060 | 1160 | 1590 | 1850 |
| Ponderosa Pine - Sugar Pine | 910 | 1000 | 1370 | 1600 |
| Red Pine | 880 | 970 | 1320 | 1540 |
| Sitka Spruce | 990 | 1090 | 1480 | 1730 |
| Southern Cypress | 1330 | 1460 | 1990 | 2320 |
| Southern Pine (Dense) | 1540 | 1690 | 2310 | 2690 |
| Southern Pine | 1320 | 1450 | 1970 | 2300 |
| Spruce - Pine - Fir | 940 | 1040 | 1410 | 1650 |
| Sweetgum and Tupelo | 1020 | 1120 | 1530 | 1780 |
| Virginia Pine - Pond Pine | 1270 | 1390 | 1900 | 2220 |
| Western Cedars[4] | 1040 | 1140 | 1520 | 1750 |
| Western Hemlock | 1240 | 1360 | 1860 | 2170 |
| Western White Pine | 930 | 1030 | 1400 | 1630 |
| White Woods (Western Woods) | 810 | 890 | 1220 | 1420 |
| West Coast Woods (Mixed Species) | 810 | 890 | 1220 | 1420 |
| Yellow Poplar | 890 | 980 | 1340 | 1560 |

1. Refer to NFPA National Design Specification sections 4.1.4 and 5.1.5 for definitions of wet and dry service conditions.
2. Applies to sawn lumber members which are at a moisture content of 19 percent or less when full design load is applied, regardless of moisture content at time of manufacture.
3. When 4 inch or thinner sawn lumber is surfaced at a moisture content of 15 percent or less and is used under dry service conditions, the values listed for glued laminated timber may be applied.
4. Values also apply when species name includes the designation "North."

By permission of NFPA

# Appendix 2

**TABLE 4A—DESIGN VALUES FOR VISUALLY GRADED STRUCTURAL LUMBER**
(Design values listed are for normal loading conditions. See other provisions in the footnotes and in the National Design Specification for adjustments of tabulated values.)

| Species and commercial grade | Size classification | Extreme fiber in bending "$F_B$" Single-member uses | Extreme fiber in bending "$F_B$" Repetitive-member uses | Tension parallel to grain "$F_T$" | Horizontal shear "$F_V$" | Compression perpendicular to grain "$F_{C\perp}$" | Compression parallel to grain "$F_C$" | Modulus of elasticity "$E$" | Grading rules agency |
|---|---|---|---|---|---|---|---|---|---|
| **COTTONWOOD** (Surfaced dry or surfaced green. Used at 19% max. m.c.) | | | | | | | | | |
| Stud | 2" to 3" thick, 2" to 4" wide | 525 | 600 | 300 | 65 | 320 | 350 | 1,000,000 | NHPMA (See footnotes 1–12) |
| Construction | 2" to 4" thick, 4" wide | 675 | 775 | 400 | 85 | 320 | 650 | 1,000,000 | |
| Standard | | 375 | 425 | 225 | 65 | 320 | 525 | 1,000,000 | |
| Utility | | 175 | 200 | 100 | 65 | 320 | 350 | 1,000,000 | |
| **DOUGLAS FIR-LARCH** (Surfaced dry or surfaced green. Used at 19% max. m.c.) | | | | | | | | | |
| Dense Select Structural | 2" to 4" thick, 2" to 4" wide | 2450 | 2800 | 1400 | 95 | 730 | 1850 | 1,900,000 | |
| Select Structural | | 2100 | 2400 | 1200 | 95 | 625 | 1600 | 1,800,000 | |
| Dense No. 1 | | 2050 | 2400 | 1200 | 95 | 730 | 1450 | 1,900,000 | |
| No. 1 | | 1750 | 2050 | 1050 | 95 | 625 | 1250 | 1,800,000 | |
| Dense No. 2 | | 1700 | 1950 | 1000 | 95 | 730 | 1150 | 1,700,000 | |
| No. 2 | | 1450 | 1650 | 850 | 95 | 625 | 1000 | 1,700,000 | |
| No. 3 | | 800 | 925 | 475 | 95 | 625 | 600 | 1,500,000 | WCLIB |
| Appearance | | 1750 | 2050 | 1050 | 95 | 625 | 1500 | 1,800,000 | WWPA |
| Stud | | 800 | 925 | 475 | 95 | 625 | 600 | 1,500,000 | |
| Construction | 2" to 4" thick, 4" wide | 1050 | 1200 | 625 | 95 | 625 | 1150 | 1,500,000 | |
| Standard | | 600 | 675 | 350 | 95 | 625 | 925 | 1,500,000 | |
| Utility | | 275 | 325 | 175 | 95 | 625 | 600 | 1,500,000 | |
| Dense Select Structural | 2" to 4" thick, 5" and wider | 2100 | 2400 | 1400 | 95 | 730 | 1650 | 1,900,000 | (See footnotes 1–12 and 20) |
| Select Structural | | 1800 | 2050 | 1200 | 95 | 625 | 1400 | 1,800,000 | |
| Dense No. 1 | | 1800 | 2050 | 1200 | 95 | 730 | 1450 | 1,900,000 | |
| No. 1 | | 1500 | 1750 | 1000 | 95 | 625 | 1250 | 1,800,000 | |
| Dense No. 2 | | 1450 | 1700 | 775 | 95 | 730 | 1250 | 1,700,000 | |
| No. 2 | | 1250 | 1450 | 650 | 95 | 625 | 1050 | 1,700,000 | |
| No. 3 | | 725 | 850 | 375 | 95 | 625 | 675 | 1,500,000 | |
| Appearance | | 1500 | 1750 | 1000 | 95 | 625 | 1500 | 1,800,000 | |
| Stud | | 725 | 850 | 375 | 95 | 625 | 675 | 1,500,000 | |
| Dense Select Structural | Beams and Stringers | 1900 | — | 1100 | 85 | 730 | 1300 | 1,700,000 | |
| Select Structural | | 1600 | — | 950 | 85 | 625 | 1100 | 1,600,000 | |
| Dense No. 1 | | 1550 | — | 775 | 85 | 730 | 1100 | 1,700,000 | |
| No. 1 | | 1300 | — | 675 | 85 | 625 | 925 | 1,600,000 | |
| No. 2 | | 875 | — | 425 | 85 | 625 | 600 | 1,300,000 | |
| Dense Select Structural | Posts and Timbers | 1750 | — | 1150 | 85 | 730 | 1350 | 1,700,000 | WCLIB (See footnotes 1–12 and 20) |
| Select Structural | | 1500 | — | 1000 | 85 | 625 | 1150 | 1,600,000 | |
| Dense No. 1 | | 1400 | — | 950 | 85 | 730 | 1200 | 1,700,000 | |
| No. 1 | | 1200 | — | 825 | 85 | 625 | 1000 | 1,600,000 | |
| No. 2 | | 750 | — | 475 | 85 | 625 | 700 | 1,300,000 | |
| Select Dex | Decking | 1750 | 2000 | — | — | 625 | — | 1,800,000 | |
| Commercial Dex | | 1450 | 1650 | — | — | 625 | — | 1,700,000 | |
| Dense Select Structural | Beams and Stringers | 1900 | — | 1250 | 85 | 730 | 1300 | 1,700,000 | |
| Select Structural | | 1600 | — | 1050 | 85 | 625 | 1100 | 1,600,000 | |
| Dense No. 1 | | 1550 | — | 1050 | 85 | 730 | 1100 | 1,700,000 | |
| No. 1 | | 1350 | — | 900 | 85 | 625 | 925 | 1,600,000 | |
| Dense No. 2 | | 1000 | — | 500 | 85 | 730 | 700 | 1,400,000 | |
| No. 2 | | 875 | — | 425 | 85 | 625 | 600 | 1,300,000 | WWPA |
| Dense Select Structural | Posts and Timbers | 1750 | — | 1150 | 85 | 730 | 1350 | 1,700,000 | |
| Select Structural | | 1500 | — | 1000 | 85 | 625 | 1150 | 1,600,000 | |
| Dense No. 1 | | 1400 | — | 950 | 85 | 730 | 1200 | 1,700,000 | |
| No. 1 | | 1200 | — | 825 | 85 | 625 | 1000 | 1,600,000 | |
| Dense No. 2 | | 800 | — | 550 | 85 | 730 | 550 | 1,400,000 | |
| No. 2 | | 700 | — | 475 | 85 | 625 | 475 | 1,300,000 | |
| Selected Decking | Decking | — | 2000 | — | — | — | — | 1,800,000 | (See footnotes 1–13 and 20) |
| Commercial Decking | | — | 1650 | — | — | — | — | 1,700,000 | |
| Selected Decking | Decking | — | 2150 | (Surfaced at 15% max. m.c. and used at 15% max. m.c.) | | | — | 1,900,000 | |
| Commercial Decking | | — | 1800 | | | | — | 1,700,000 | |

By permission of NFPA

## Appendix 2 (cont'd)

### DESIGN VALUES FOR VISUALLY GRADED STRUCTURAL LUMBER

(Design values listed are for normal loading conditions. See other provisions in the footnotes and in the National Design Specification for adjustments of tabulated values.)

| Species and commercial grade | Size classification | Extreme fiber in bending "$F_B$" Single-member uses | Extreme fiber in bending "$F_B$" Repetitive-member uses | Tension parallel to grain "$F_T$" | Horizontal shear "$F_V$" | Compression perpendicular to grain "$F_{C\perp}$" | Compression parallel to grain "$F_C$" | Modulus of elasticity "$E$" | Grading rules agency |
|---|---|---|---|---|---|---|---|---|---|
| **SOUTHERN PINE** (Surfaced at 15% maximum moisture content, K.D.-15. Used at 15% max. m.c.) | | | | | | | | | |
| Select Structural | | 2150 | 2500 | 1250 | 105 | 565 | 1800 | 1,800,000 | |
| Dense Select Structural | | 2500 | 2900 | 1500 | 105 | 660 | 2100 | 1,900,000 | |
| No. 1 | | 1850 | 2100 | 1050 | 105 | 565 | 1450 | 1,800,000 | |
| No. 1 Dense | 2" to 4" | 2150 | 2450 | 1250 | 105 | 660 | 1700 | 1,900,000 | |
| No. 2 | thick | 1550 | 1750 | 900 | 95 | 565 | 1150 | 1,600,000 | |
| No. 2 Dense | 2" to 4" | 1800 | 2050 | 1050 | 95 | 660 | 1350 | 1,700,000 | |
| No. 3 | wide | 850 | 975 | 500 | 95 | 565 | 675 | 1,500,000 | |
| No. 3 Dense | | 1000 | 1150 | 575 | 95 | 660 | 800 | 1,500,000 | |
| Stud | | 850 | 975 | 500 | 95 | 565 | 675 | 1,500,000 | |
| Construction | 2" to 4" | 1100 | 1250 | 650 | 105 | 565 | 1300 | 1,500,000 | |
| Standard | thick | 625 | 725 | 375 | 95 | 565 | 1050 | 1,500,000 | |
| Utility | 4" wide | 275 | 300 | 175 | 95 | 565 | 675 | 1,500,000 | SPIB |
| Select Structural | | 1850 | 2150 | 1200 | 95 | 565 | 1600 | 1,800,000 | |
| Dense Select Structural | | 2200 | 2500 | 1450 | 95 | 660 | 1850 | 1,900,000 | |
| No. 1 | | 1600 | 1850 | 1050 | 95 | 565 | 1450 | 1,800,000 | |
| No. 1 Dense | 2" to 4" | 1850 | 2150 | 1250 | 95 | 660 | 1700 | 1,900,000 | (See footnotes 1, 3, 4, 5, 6, 11, 12, 17, 18, 19, and 20) |
| No. 2 | thick | 1300 | 1500 | 675 (See Footnote 3) | 95 | 565 | 1200 | 1,600,000 | |
| No. 2 Dense | 5" and wider | 1550 | 1750 | 800 | 95 | 660 | 1400 | 1,700,000 | |
| No. 3 | | 750 | 875 | 400 | 95 | 565 | 725 | 1,500,000 | |
| No. 3 Dense | | 875 | 1000 | 450 | 95 | 660 | 850 | 1,500,000 | |
| Stud | | 800 | 900 | 400 | 95 | 565 | 725 | 1,500,000 | |
| Dense Standard Decking | 2" to 4" | 2150 | 2450 | — | — | 660 | — | 1,900,000 | |
| Select Decking | thick | 1550 | 1750 | — | — | 565 | — | 1,600,000 | |
| Dense Select Decking | 2" and | 1800 | 2050 | — | — | 660 | — | 1,700,000 | |
| Commercial Decking | wider | 1550 | 1750 | — | — | 565 | — | 1,600,000 | |
| Dense Commercial Decking | Decking | 1800 | 2050 | — | — | 660 | — | 1,700,000 | |
| Dense Structural 86 | 2" to 4" | 2800 | 3250 | 1900 | 165 | 660 | 2300 | 1,900,000 | |
| Dense Structural 72 | thick | 2400 | 2750 | 1600 | 135 | 660 | 1950 | 1,900,000 | |
| Dense Structural 65 | | 2150 | 2450 | 1450 | 125 | 660 | 1750 | 1,900,000 | |
| **SOUTHERN PINE** (Surfaced dry. Used at 19% max. m.c.) | | | | | | | | | |
| Select Structural | | 2000 | 2300 | 1150 | 100 | 565 | 1550 | 1,700,000 | |
| Dense Select Structural | | 2350 | 2700 | 1350 | 100 | 660 | 1800 | 1,800,000 | |
| No. 1 | | 1700 | 1950 | 1000 | 100 | 565 | 1250 | 1,700,000 | |
| No. 1 Dense | 2" to 4" | 2000 | 2300 | 1150 | 100 | 660 | 1450 | 1,800,000 | |
| No. 2 | thick | 1400 | 1650 | 825 | 90 | 565 | 975 | 1,600,000 | |
| No. 2 Dense | 2" to 4" | 1650 | 1900 | 975 | 90 | 660 | 1150 | 1,600,000 | |
| No. 3 | wide | 775 | 900 | 450 | 90 | 565 | 575 | 1,400,000 | |
| No. 3 Dense | | 925 | 1050 | 525 | 90 | 660 | 675 | 1,500,000 | |
| Stud | | 775 | 900 | 450 | 90 | 565 | 575 | 1,400,000 | |
| Construction | 2" to 4" | 1000 | 1150 | 600 | 100 | 565 | 1100 | 1,400,000 | |
| Standard | thick | 575 | 675 | 350 | 90 | 565 | 900 | 1,400,000 | |
| Utility | 4" wide | 275 | 300 | 150 | 90 | 565 | 575 | 1,400,000 | |
| Select Structural | | 1750 | 2000 | 1150 | 90 | 565 | 1350 | 1,700,000 | |
| Dense Select Structural | | 2050 | 2350 | 1300 | 90 | 660 | 1600 | 1,800,000 | |
| No. 1 | | 1450 | 1700 | 975 | 90 | 565 | 1250 | 1,700,000 | SPIB |
| No. 1 Dense | 2" to 4" | 1700 | 2000 | 1150 | 90 | 660 | 1450 | 1,800,000 | (See footnotes 1,3,4,5,6,11,12, 17,18,19 and 20) |
| No. 2 | thick | 1200 | 1400 | 625 | 90 | 565 | 1000 | 1,600,000 | |
| No. 2 Dense | 5" and wider | 1400 | 1650 | 725 | 90 | 660 | 1200 | 1,600,000 | |
| No. 3 | | 700 | 800 | 350 | 90 | 565 | 625 | 1,400,000 | |
| No. 3 Dense | | 825 | 925 | 425 | 90 | 660 | 725 | 1,500,000 | |
| Stud | | 725 | 850 | 350 | 90 | 565 | 625 | 1,400,000 | |
| Dense Standard Decking | 2" to 4" | 2000 | 2300 | — | — | 660 | — | 1,800,000 | |
| Select Decking | thick | 1400 | 1650 | — | — | 565 | — | 1,600,000 | |
| Dense Select Decking | 2" and | 1650 | 1900 | — | — | 660 | — | 1,600,000 | |
| Commercial Decking | wider | 1400 | 1650 | — | — | 565 | — | 1,600,000 | |
| Dense Commercial Decking | Decking | 1650 | 1900 | — | — | 660 | — | 1,600,000 | |
| Dense Structural 86 | 2" to 4" | 2600 | 3000 | 1750 | 155 | 660 | 2000 | 1,800,000 | |
| Dense Structural 72 | thick | 2200 | 2550 | 1450 | 130 | 660 | 1650 | 1,800,000 | |
| Dense Structural 65 | | 2000 | 2300 | 1300 | 115 | 660 | 1500 | 1,800,000 | |

By permission of NFPA

# Appendix 2 (cont'd)

## DESIGN VALUES FOR VISUALLY GRADED STRUCTURAL LUMBER

(Design values listed are for normal loading conditions. See other provisions in the footnotes and in the National Design Specification for adjustments of tabulated values.)

| Species and commercial grade | Size classification | Design values in pounds per square inch ||||||| Grading rules agency |
|---|---|---|---|---|---|---|---|---|---|
| | | Extreme fiber in bending "$F_B$" || Tension parallel to grain "$F_T$" | Horizontal shear "$F_V$" | Compression perpendicular to grain "$F_{C\perp}$" | Compression parallel to grain "$F_C$" | Modulus of elasticity "$E$" | |
| | | Single-member uses | Repetitive-member uses | | | | | | |
| **SOUTHERN PINE** (Surfaced green. Used any condition) | | | | | | | | | |
| Select Structural | | 1600 | 1850 | 925 | 95 | 375 | 1050 | 1,500,000 | |
| Dense Select Structural | | 1850 | 2150 | 1100 | 95 | 440 | 1200 | 1,600,000 | |
| No. 1 | | 1350 | 1550 | 800 | 95 | 375 | 825 | 1,500,000 | |
| No. 1 Dense | 2½" to 4" | 1600 | 1800 | 925 | 95 | 440 | 950 | 1,600,000 | |
| No. 2 | thick | 1150 | 1300 | 675 | 85 | 375 | 650 | 1,400,000 | |
| No. 2 Dense | 2½" to 4" | 1350 | 1500 | 775 | 85 | 440 | 750 | 1,400,000 | |
| No. 3 | wide | 625 | 725 | 375 | 85 | 375 | 400 | 1,200,000 | |
| No. 3 Dense | | 725 | 850 | 425 | 85 | 440 | 450 | 1,300,000 | |
| Stud | | 625 | 725 | 375 | 85 | 375 | 400 | 1,200,000 | |
| Construction | 2½" to 4" | 825 | 925 | 475 | 95 | 375 | 725 | 1,200,000 | |
| Standard | thick | 475 | 525 | 275 | 85 | 375 | 600 | 1,200,000 | |
| Utility | 4" wide | 200 | 250 | 125 | 85 | 375 | 400 | 1,200,000 | |
| Select Structural | | 1400 | 1600 | 900 | 85 | 375 | 900 | 1,500,000 | |
| Dense Select Structural | | 1600 | 1850 | 1050 | 85 | 440 | 1050 | 1,600,000 | |
| No. 1 | | 1200 | 1350 | 775 (See Footnote 3) | 85 | 375 | 825 | 1,500,000 | |
| No. 1 Dense | 2½" to 4" | 1400 | 1600 | 925 (See Footnote 3) | 85 | 440 | 950 | 1,600,000 | SPIB |
| No. 2 | thick | 975 | 1100 | 500 (See Footnote 3) | 85 | 375 | 675 | 1,400,000 | |
| No. 2 Dense | 5" and wider | 1150 | 1300 | 600 (See Footnote 3) | 85 | 440 | 800 | 1,400,000 | (See footnotes |
| No. 3 | | 550 | 650 | 300 (See Footnote 3) | 85 | 375 | 425 | 1,200,000 | 1,3,4,5,6,11,12, |
| No. 3 Dense | | 650 | 750 | 350 | 85 | 440 | 475 | 1,300,000 | 17,18,19 and 20) |
| Stud | | 575 | 675 | 300 | 85 | 375 | 425 | 1,200,000 | |
| Dense Standard Decking | 2½" to 4" | 1600 | 1800 | — | — | 440 | — | 1,600,000 | |
| Select Decking | thick | 1150 | 1300 | — | — | 375 | — | 1,400,000 | |
| Dense Select Decking | 2" and | 1350 | 1500 | — | — | 440 | — | 1,400,000 | |
| Commercial Decking | wider | 1150 | 1300 | — | — | 375 | — | 1,400,000 | |
| Dense Commercial Decking | Decking | 1350 | 1500 | — | — | 440 | — | 1,400,000 | |
| No. 1 SR | | 1350 | — | 875 | 110 | 375 | 775 | 1,500,000 | |
| No. 1 Dense SR | 5" and | 1550 | — | 1050 | 110 | 440 | 925 | 1,600,000 | |
| No. 2 SR | thicker | 1100 | — | 725 | 95 | 375 | 625 | 1,400,000 | |
| No. 2 Dense SR | | 1250 | — | 850 | 95 | 440 | 725 | 1,400,000 | |
| Dense Structural 86 | 2½" and | 2100 | 2400 | 1400 | 145 | 440 | 1300 | 1,600,000 | |
| Dense Structural 72 | thicker | 1750 | 2050 | 1200 | 120 | 440 | 1100 | 1,600,000 | |
| Dense Structural 65 | | 1600 | 1800 | 1050 | 110 | 440 | 1000 | 1,600,000 | |

By permission of NFPA

# Timber Design

## Appendix 2 (cont'd)
### Footnotes
#### Applicable to Visually Graded Structural Lumber

1. Following is a list of agencies certified by the American Lumber Standards Committee Board of Review (as of 1982) for inspection and grading of untreated lumber under the rules indicated. For the most up-to-date list of certified agencies, write to:

American Lumber Standards Committee
P.O. Box 210
Germantown, Maryland 20874

| Rules Writing Agencies | Rules for which grading authorized |
|---|---|
| Northeastern Lumber Manufacturers Association (NELMA) | NELMA, NLGA |
| 4 Fundy Road, Falmouth, Maine 04105 | |
| Northern Hardwood and Pine Manufacturers Association (serviced by NELMA) | NHPMA, WCLIB, WWPA, NLGA |
| 4 Fundy Road, Falmouth, Maine 04105 | |
| Redwood Inspection Service (RIS) | RIS, WCLIB, WWPA |
| 591 Redwood Highway, Suite 3100, Mill Valley, California 94941 | |
| Southern Pine Inspection Bureau (SPIB) | SPIB, NELMA |
| 4709 Scenic Highway, Pensacola, Florida 32504 | |
| West Coast Lumber Inspection Bureau (WCLIB) | WCLIB, RIS, WWPA, NLGA |
| 6980 SW Varnes Rd., PO Box 23145, Portland, OR 97223 | |
| Western Wood Products Association (WWPA) | WWPA, WCLIB, NLGA, RIS |
| 1500 Yeon Building, Portland, Oregon 97204 | |
| National Lumber Grades Authority (NLGA) | |
| P.O. Box 97 Ganges, B.C., Canada VOS 1EO | |

| Non-Rules Writing Agencies | |
|---|---|
| California Lumber Inspection Service | RIS, WCLIB, WWPA, NLGA |
| Pacific Lumber Inspection Bureau, Inc. | RIS, WCLIB, WWPA, NLGA |
| Timber Products Inspection | RIS, SPIB, WCLIB, WWPA, NHPMA, NELMA, NLGA |
| Alberta Forest Products Association | NLGA |
| Canadian Lumbermans Association | NLGA |
| Cariboo Lumber Manufacturers Association | NLGA |
| Central Forest Products Association | NLGA |
| Council of Forest Industries of British Columbia | NLGA |
| Interior Lumber Manufacturers Association | NLGA |
| MacDonald Inspection | NLGA |
| Maritime Lumber Bureau | NLGA |
| Ontario Lumber Manufacturers Association | NLGA |
| Pacific Lumber Inspection Bureau | NLGA |
| Quebec Lumber Manufacturers Association | NLGA |

2. The design values herein are applicable to lumber that will be used under dry conditions such as in most covered structures. For 2" to 4" thick lumber the DRY surfaced size shall be used. In calculating design values, the natural gain in strength and stiffness that occurs as lumber dries has been taken into consideration as well as the reduction in size that occurs when unseasoned lumber shrinks. The gain in load carrying capacity due to increased strength and stiffness resulting from drying more than offsets the design effect of size reductions due to shrinkage. For 5" and thicker lumber, the surfaced sizes also may be used because design values have been adjusted to compensate for any loss in size by shrinkage which may occur.

3. Tabulated tension parallel to grian values for all species for 5" and wider, 2" to 4" thick (and 2½" to 4" thick) size classifications apply to 5" and 6" widths only, for grades of Select Structural, No. 1, No. 2, No. 3, Appearance and Stud, (including dense grades). For lumber wider than 6" in these grades, the tabulated "$F_T$" values shall be multiplied by the following factors:

| Grade (2" to 4" thick, 5" and wider) (2½" to 4" thick, 5" and wider) (Includes "Dense" grades) | Multiply tabulated "$F_T$" values by | | |
|---|---|---|---|
| | 5" & 6" wide | 8" wide | 10" and wider |
| Select Structural | 1.00 | 0.90 | 0.80 |
| No. 1, No. 2, No. 3 and Appearance | 1.00 | 0.80 | 0.60 |
| Stud | 1.00 | — | — |

4. Design values for all species of Stud grade in 5" and wider size classifications apply to 5" and 6" widths only.

5. Values for "$F_B$", "$F_T$", and "$F_C$" for all species of the grades of Construction, Standard and Utility apply only to 4" widths. Design values for 2" and 3" widths of these grades are available from the grading rules agencies (see Note 1).

6. The values in Table 4A for dimension lumber 2" to 4" in thickness are based on edgewise use. When such lumber is used flatwise, the design values for extreme fiber in bending for all species may be multiplied by the following factors:

| Width | Dimension lumber used flatwise Thickness | | |
|---|---|---|---|
| | 2" | 3" | 4" |
| 2" to 4" | 1.10 | 1.04 | 1.00 |
| 5" and wider | 1.22 | 1.16 | 1.11 |

7. The design values in Table 4A for extreme fiber in bending for decking may be increased by 10 percent for 2" thick decking and by 4 percent for 3" thick decking. (Not applicable to California Redwood.)

8. When 2" to 4" thick lumber is manufactured at a maximum moisture content of 15 percent and used in a condition where the moisture content does not exceed 15 percent, the design values for surfaced dry or surfaced green lumber shown in Table 4A may be multiplied by the following factors. (For Southern Pine and Virginia Pine–Pond Pine use tabulated design values without adjustment):

| 2" to 4" thick lumber manufactured and used at 15 percent maximum moisture content (m.c. 15) | | | | | |
|---|---|---|---|---|---|
| Extreme fiber in bending "$F_B$" | Tension parallel to grain "$F_T$" | Horizontal shear "$F_V$" | Compression perpendicular to grain "$F_{C\perp}$" | Compression* parallel to grain "$F_C$" | Modulus* of elasticity "E" |
| 1.08 | 1.08 | 1.05 | 1.00 | 1.17 | 1.05 |
| | | | *For Redwood use only | 1.15 | 1.04 |

### By permission or NFPA
PROFESSIONAL ENGINEERING REGISTRATION PROGRAM • P.O. Box 911, San Carlos, CA 94070

## Appendix 2 (cont'd)

9. When 2" to 4" thick lumber is designed for use where the moisture content will exceed 19 percent for an extended period of time, the design values shown herein shall be multiplied by the following factors, except that for Southern Pine and Virginia Pine–Pond Pine footnote 18 applies:

| 2" to 4" thick lumber used where moisture content will exceed 19% | | | | | |
|---|---|---|---|---|---|
| Extreme fiber in bending "$F_B$" | Tension parallel to grain "$F_T$" | Horizontal shear "$F_V$" | Compression perpendicular to grain "$F_{C\perp}$" | Compression parallel to grain "$F_C$" | Modulus of elasticity "$E$" |
| 0.86 | 0.84 | 0.97 | 0.67 | 0.70 | 0.97 |

10. When lumber 5" and thicker is designed for use where the moisture content will exceed 19 percent for an extended period of time, the design values shown in Table 4A (except those for Southern Pine and Virginia Pine–Pond Pine) shall be multiplied by the following factors:

| 5" and thicker lumber used where moisture content will exceed 19% | | | | | |
|---|---|---|---|---|---|
| Extreme fiber in bending "$F_B$" | Tension parallel to grain "$F_T$" | Horizontal shear "$F_V$" | Compression perpendicular to grain "$F_{C\perp}$" | Compression parallel to grain "$F_C$" | Modulus of elasticity "$E$" |
| 1.00 | 1.00 | 1.00 | 0.67 | 0.91 | 1.00 |

11. Specific horizontal shear values may be established by use of the following table when length of split, or size of check or shake is known and no increase in them is anticipated. For California Redwood, Southern Pine, Virginia Pine–Pond Pine, or Yellow-Poplar, the provisions in this Footnote apply only to the following $F_V$ Values: 80 psi, California Redwood; 95 psi, Southern Pine (KD-15); 90 psi, Southern Pine (S-Dry); 85 psi, Southern Pine (S-Green); 95 psi, Virginia Pine–Pond Pine (KD-15); 90 psi, Virginia Pine–Pond Pine (S-Dry); 85 psi, Virginia Pine–Pond Pine (S-Green); and 75 psi, Yellow-Poplar.

| Shear Stress Modification Factor | | | | | |
|---|---|---|---|---|---|
| Length of split on wide face of 2" lumber (nominal): | Multiply tabulated "$F_V$" value by: | Length of split on wide face of 3" and thicker lumber (nominal): | Multiply tabulated "$F_V$" value by: | Size of shake* in 3" and thicker lumber (nominal): | Multiply tabulated "$F_V$" value by: |
| no split | 2.00 | no split | 2.00 | no shake | 2.00 |
| ½ × wide face | 1.67 | ½ × narrow face | 1.67 | 1/6 × narrow face | 1.67 |
| ¾ × wide face | 1.50 | 1 × narrow face | 1.33 | 1/3 narrow face | 1.33 |
| 1 × wide face | 1.33 | 1½ × narrow face or more | 1.00 | ½ × narrow face or more | 1.00 |
| 1½ × wide face or more | 1.00 | | | *Shake is measured at the end between lines enclosing the shake and parallel to the wide face. | |

12. Stress rated boards of nominal 1", 1¼" and 1½" thickness, 2" and wider, of most species, are permitted the design values shown for Select Structural, No. 1, No. 2, No. 3, Construction, Standard, Utility, Appearance, Clear Heart Structural and Clear Structural grades as shown in the 2" to 4" thick categories herein, when graded in accordance with the stress rated board provisions in the applicable grading rules. Information on stress rated board grades applicable to the various species is available from the respective grading rules agencies. Information on additional design values may also be available from the respective grading agencies.

13. When Decking graded to WWPA rules is surfaced at 15 percent maximum moisture content and used where the moisture content will exceed 15 percent for an extended period of time, the tabulated design values for Decking surfaced at 15 percent maximum moisture content shall be multiplied by the following factors: Extreme Fiber in Bending "$F_B$", 0.79; Modulus of Elasticity "$E$", 0.92.

14. To obtain a recommended design value for Spruce Pine, multiply the appropriate design value for Virginia Pine–Pond Pine by the corresponding conversion factor shown below and round to the nearest 100,000 psi for modulus of elasticity; to the next lower multiple of 5 psi for horizontal shear and compression perpendicular to grain; to the next lower multiple of 50 psi for bending, tension parallel to grain and compression parallel to grain if 1000 psi or greater, 25 psi otherwise.

| Conversion Factors for Determining Design Values for Spruce Pine | | | | | | | |
|---|---|---|---|---|---|---|---|
| Design Category | Extreme fiber in bending "$F_B$" | | Tension parallel to grain "$F_T$" | Horizontal Shear "$F_V$" | Compression perpendicular to grain "$F_{C\perp}$" | Compression parallel to grain "$F_C$" | Modulus of elasticity "$E$" |
| | Single member uses | Repetitive member uses | | | | | |
| Conversion Factor | .784 | .784 | .784 | .766 | .965 | .682 | .807 |

15. National Lumber Grades Authority is the Canadian rules writing agency responsible for preparation, maintenance and dissemination of a uniform softwood lumber grading rule for all Canadian species.

16. For species graded to NLGA rules, values shown in Table 4A for Select Structural, No. 1, No. 2, No. 3 and Stud grades are not applicable to 3" × 4" and 4" × 4" sizes.

17. Repetitive member design values for extreme fiber in bending for Southern Pine grades of Dense Structural 86, 72 and 65 apply to 2" to 4" thicknesses only.

18. When 2" to 4" thick Southern Pine or Virginia Pine–Pond Pine lumber is surfaced dry or at 15 percent maximum moisture content (KD-15) and is designed for use where the moisture content will exceed 19 percent for an extended period of time, the design values in Table 4A for the corresponding grades of 2½" to 4" thick surfaced green Southern Pine lumber shall be used. The net green size may be used in such designs.

19. When 2" to 4" thick Southern Pine or Virginia Pine–Pond Pine lumber is surfaced dry or at 15 percent maximum moisture content (KD-15) and is designed for use under dry conditions, such as in most covered structures, the net DRY size shall be used in design. For other sizes and conditions of use, the net green size may be used in design.

20. When the depth of a beam, stringer, post, timber or other rectangular sawn lumber member 5" or thicker exceeds 12", the design value for the extreme fiber in bending, $F_B$, shall be multiplied by the size factor, $C_F$, as determined by the following formula:

$$C_F = \left(\frac{12}{d}\right)^{1/9}$$

By permission of NFPA

## Appendix 2 (cont'd)

**DESIGN VALUES FOR STRUCTURAL GLUED LAMINATED SOFTWOOD TIMBERS: MEMBERS STRESSED PRIMARILY IN BENDING**[1,2,3,4,6,14]

Design values are for normal load duration and dry conditions of use. See footnotes, and other provisions in the National Design Specification for Wood Construction, for adjustments of calculated values.

Design values in pounds per square inch

| Combination Symbol[6] | Species Outer Laminations/ Core Laminations[7] | BENDING ABOUT X-X AXIS ||||||| BENDING ABOUT Y-Y AXIS ||||| AXIALLY LOADED |||
|---|---|---|---|---|---|---|---|---|---|---|---|---|---|---|---|---|
| | | Loaded Perpendicular to Wide Faces of Laminations ||| Compression Perp. to Grain || Horizontal Shear[12] | Modulus of Elasticity | Loaded Parallel to the Wide Faces of Laminations |||| | | | |
| | | Extreme Fiber in Bending[5] || Tension Face[11,12] | Compression Face[11,12] | | | Extreme Fiber in Bending[5,15] | Compression Perpendicular to Grain Side faces | Horizontal Shear | Horizontal Shear (For Members With Multiple Piece Laminations Which Are Not Edge Glued)[16] | Modulus of Elasticity | Tension Parallel to Grain | Compression Parallel to Grain | Modulus of Elasticity |
| | | Tension Zone Stressed in Tension | Compression Zone Stressed[8] in Tension | | | | | | | | | | | | |
| | | $F_{bxx}$ | $F_{bxx}$ | $F_{c \perp xx}$ | $F_{c \perp xx}$ | $F_{vxx}$ | $E_{xx}$ | $F_{byy}$ | $F_{c \perp yy}$ | $F_{vyy}$ | $F_{vyy}$ | $E_{yy}$ | $F_t$ | $F_c$ | $E$ |
| **VISUALLY GRADED WESTERN SPECIES** ||||||||||||||||
| 16F-V1 | DF/WW | 1600 | 800 | 560[11,12] | 560[12] | 140 | 1,300,000 | 950 | 255 | 130[17] | 65[17] | 1,100,000 | 675 | 975 | 1,100,000 |
| 16F-V2 | HF/HF | 1600 | 800 | 500[12] | 375[12] | 155 | 1,400,000 | 1250 | 375 | 135 | 70 | 1,300,000 | 875 | 1300 | 1,300,000 |
| 16F-V3 | DF/DF | 1600 | 800 | 560[11,12] | 560[12] | 165 | 1,500,000 | 1450 | 560 | 145 | 70 | 1,500,000 | 950 | 1550 | 1,500,000 |
| 16F-V4[9] | DF/N3WW | 1600 | 800 | 650 | 560[12] | 90[12] | 1,500,000 | 900 | 255 | 130[17] | 65[17] | 1,300,000 | 650 | 600 | 1,300,000 |
| 16F-V5[9] | DF/N3DF | 1600 | 800 | 650 | 560[12] | 90[12] | 1,600,000 | 1000 | 470 | 135 | 70 | 1,500,000 | 750 | 875 | 1,500,000 |
| 16F-V6[10] | DF/DF | 1600 | 1600 | 560[11,12] | 560[12] | 165 | 1,500,000 | 1450 | 560 | 145 | 75 | 1,400,000 | 950 | 1550 | 1,500,000 |
| 16F-V7[10] | HF/HF | 1600 | 1600 | 375[12] | 375[12] | 155 | 1,400,000 | 1200 | 375 | 135 | 70 | 1,300,000 | 850 | 1350 | 1,300,000 |
| 16F-V8 | DFS/DFS | 1600 | 800 | 650 | 500 | 165 | 1,200,000 | 1200 | 500 | 145 | 75 | 1,100,000 | 825 | 1350 | 1,100,000 |
| 20F-V1 | DF/WW | 2000 | 1000 | 650 | 560[12] | 140 | 1,400,000 | 1000 | 255 | 130[17] | 65[17] | 1,200,000 | 750 | 1000 | 1,200,000 |
| 20F-V2 | HF/HF | 2000 | 1000 | 500[12] | 375[12] | 155 | 1,500,000 | 1200 | 375 | 135 | 70 | 1,400,000 | 950 | 1350 | 1,400,000 |
| 20F-V3 | DF/DF | 2000 | 1000 | 650 | 560[12] | 165 | 1,600,000 | 1450 | 560 | 145 | 75 | 1,500,000 | 1000 | 1550 | 1,500,000 |
| 20F-V4 | DF/DF | 2000 | 1000 | 590[11,12] | 560[12] | 165 | 1,600,000 | 1450 | 560 | 145 | 75 | 1,600,000 | 1000 | 1550 | 1,600,000 |
| 20F-V5[9] | DF/N3WW | 2000 | 1000 | 650 | 560[12] | 90[12] | 1,600,000 | 1000 | 255 | 135[17] | 70[17] | 1,300,000 | 750 | 725 | 1,300,000 |
| 20F-V6[9] | DF/N3DF | 2000 | 1000 | 650 | 560[12] | 90[12] | 1,600,000 | 1000 | 470 | 135 | 70 | 1,500,000 | 775 | 900 | 1,500,000 |
| 20F-V7[10] | DF/DF | 2000 | 2000 | 650 | 650 | 165 | 1,600,000 | 1450 | 560 | 145 | 75 | 1,600,000 | 1000 | 1600 | 1,600,000 |
| 20F-V8[10] | DF/DF | 2000 | 2000 | 590[11,12] | 590[11,12] | 165 | 1,700,000 | 1450 | 560 | 145 | 75 | 1,600,000 | 1000 | 1600 | 1,600,000 |
| 20F-V9[10] | HF/HF | 2000 | 2000 | 500[12] | 500[12] | 155 | 1,500,000 | 1400 | 375 | 135 | 70 | 1,400,000 | 975 | 1400 | 1,400,000 |
| 20F-V10 | DF/HF | 2000 | 1000 | 560 | 560 | 155 | 1,500,000 | 1300 | 375 | 135 | 70 | 1,400,000 | 950 | 1500 | 1,400,000 |
| 20F-V11 | DFS/DFS | 2000 | 1000 | 650 | 500 | 165 | 1,300,000 | 1400 | 500 | 145 | 75 | 1,100,000 | 900 | 1400 | 1,100,000 |
| 22F-V1 | DF/WW | 2200 | 1100 | 650 | 560[12] | 140 | 1,600,000 | 1050 | 255 | 130[17] | 65[17] | 1,300,000 | 850 | 1100 | 1,300,000 |
| 22F-V2 | HF/HF | 2200 | 1100 | 500[12] | 500[12] | 155 | 1,500,000 | 1250 | 375 | 135 | 70 | 1,400,000 | 950 | 1350 | 1,400,000 |
| 22F-V3 | DF/DF | 2200 | 1100 | 650 | 560[12] | 165 | 1,600,000 | 1450 | 560 | 145 | 75 | 1,600,000 | 1000 | 1500 | 1,600,000 |
| 22F-V4 | DF/DF | 2200 | 1100 | 590[11,12] | 560[12] | 165 | 1,700,000 | 1450 | 560 | 145 | 75 | 1,600,000 | 1000 | 1550 | 1,600,000 |
| 22F-V5[9] | DF/N3WW | 2200 | 1100 | 650 | 560[12] | 90[12] | 1,600,000 | 1100 | 255 | 135[17] | 75[17] | 1,400,000 | 800 | 725 | 1,400,000 |
| 22F-V6[9] | DF/N3DF | 2200 | 1100 | 650 | 560[12] | 90[12] | 1,700,000 | 1250 | 470 | 135 | 75 | 1,600,000 | 900 | 925 | 1,500,000 |
| 22F-V7[10] | DF/DF | 2200 | 2200 | 650 | 650 | 165 | 1,800,000 | 1450 | 560 | 145 | 75 | 1,600,000 | 1100 | 1650 | 1,600,000 |
| 22F-V8[10] | DF/DF | 2200 | 2200 | 590[11,12] | 590[11,12] | 165 | 1,700,000 | 1450 | 560 | 145 | 75 | 1,600,000 | 1050 | 1650 | 1,800,000 |
| 22F-V9[10] | HF/HF | 2200 | 2200 | 500[12] | 500[12] | 155 | 1,500,000 | 1250 | 375 | 135 | 70 | 1,400,000 | 975 | 1400 | 1,400,000 |
| 22F-V10 | DF/DFS | 2200 | 1100 | 650 | 560[12] | 165 | 1,600,000 | 1600 | 500 | 145 | 75 | 1,300,000 | 1000 | 1400 | 1,300,000 |
| 24F-V1 | DF/WW | 2400 | 1200 | 650 | 650 | 140 | 1,700,000 | 1250 | 255 | 135[17] | 70[17] | 1,400,000 | 1000 | 1300 | 1,400,000 |
| 24F-V2 | HF/HF | 2400 | 1200 | 500[12] | 500[12] | 155 | 1,500,000 | 1250 | 375 | 135 | 70 | 1,400,000 | 950 | 1300 | 1,400,000 |
| 24F-V3 | DF/DF | 2400 | 1200 | 650 | 560[12] | 165 | 1,800,000 | 1500 | 560 | 145 | 75 | 1,600,000 | 1100 | 1600 | 1,600,000 |
| 24F-V4 | DF/DF | 2400 | 1200 | 650 | 650 | 165 | 1,700,000 | 1500 | 560 | 145 | 75 | 1,600,000 | 1150 | 1650 | 1,600,000 |
| 24F-V5 | DF/HF | 2400 | 1200 | 650 | 650 | 155 | 1,700,000 | 1350 | 375 | 140 | 70 | 1,500,000 | 1100 | 1450 | 1,500,000 |
| 24F-V6[9] | DF/N3WW | 2400 | 1200 | 650 | 560[12] | 90[12] | 1,700,000 | 1200 | 255 | 140[17] | 70[17] | 1,500,000 | 950 | 800 | 1,500,000 |
| 24F-V7[9] | DF/N3DF | 2400 | 1200 | 650 | 560[12] | 90[12] | 1,700,000 | 1250 | 470 | 135 | 70 | 1,600,000 | 900 | 950 | 1,500,000 |
| 24F-V8[10] | DF/DF | 2400 | 2400 | 650 | 650 | 165 | 1,800,000 | 1450 | 560 | 145 | 75 | 1,600,000 | 1100 | 1650 | 1,600,000 |
| 24F-V9[10] | HF/HF | 2400 | 2400 | 500[12] | 500[12] | 155 | 1,500,000 | 1500 | 375 | 135 | 70 | 1,400,000 | 1000 | 1450 | 1,400,000 |
| 24F-V10[10] | DF/HF | 2400 | 2400 | 650 | 650 | 155 | 1,800,000 | 1400 | 375 | 140 | 70 | 1,600,000 | 1150 | 1600 | 1,600,000 |
| 24F-V11 | DF/DFS | 2400 | 1200 | 650 | 560[12] | 165 | 1,700,000 | 1600 | 500 | 145 | 75 | 1,400,000 | 1150 | 1700 | 1,400,000 |
| Wet-use factors | | 0.8 | 0.8 | 0.53 | 0.53 | 0.876 | 0.833 | 0.8 | 0.53 | 0.875 | 0.875 | 0.833 | 0.8 | 0.73 | 0.833 |

By permission of NFPA

## Appendix 2 (cont'd)

**DESIGN VALUES FOR STRUCTURAL GLUED LAMINATED SOFTWOOD TIMBERS: MEMBERS STRESSED PRIMARILY IN BENDING**[1,2,3,4,6,14]

Design values are for normal load duration and dry conditions of use. See footnotes, and other provisions in the National Design Specification for Wood Construction, for adjustments of calculated values.

Design values in pounds per square inch

| Combination Symbol[6] | Species Outer Laminations/ Core Laminations[7] | BENDING ABOUT X-X AXIS — Extreme Fiber in Bending[5] — Tension Zone Stressed in Tension $F_{bxx}$ | Compression Zone Stressed[8] in Tension $F_{bxx}$ | Compression Perp. to Grain — Tension Face[11,12] $F_{c\perp xx}$ | Compression Face[11,12] $F_{c\perp xx}$ | Horizontal Shear[12] $F_{vxx}$ | Modulus of Elasticity $E_{xx}$ | BENDING ABOUT Y-Y AXIS — Extreme Fiber in Bending[5,15] $F_{byy}$ | Compression Perpendicular to Grain Side faces $F_{c\perp yy}$ | Horizontal Shear $F_{vyy}$ | Horizontal Shear (For Members With Multiple Piece Laminations Which Are Not Edge Glued)[16] $F'_{vyy}$ | Modulus of Elasticity $E_{yy}$ | AXIALLY LOADED — Tension Parallel to Grain $F_t$ | Compression Parallel to Grain $F_c$ | Modulus of Elasticity $E$ |
|---|---|---|---|---|---|---|---|---|---|---|---|---|---|---|---|
| **VISUALLY GRADED SOUTHERN PINE** | | | | | | | | | | | | | | | |
| 16F-V1 | SP/SP | 1600 | 800 | 560[11,12] | 560[12] | 200 | 1,400,000 | 1450 | 560 | 175 | 90 | 1,300,000 | 950 | 1450 | 1,300,000 |
| 16F-V2 | SP/SP | 1600 | 800 | 560[11,12] | 560[12] | 200 | 1,400,000 | 1600 | 560 | 175 | 90 | 1,400,000 | 1000 | 1550 | 1,400,000 |
| 16F-V3 | SP/SP | 1600 | 800 | 650 | 650 | 200 | 1,400,000 | 1450 | 560 | 175 | 90 | 1,300,000 | 975 | 1450 | 1,300,000 |
| 16F-V4[9] | SP/SP | 1600 | 800 | 560[11,12] | 560[12] | 90[12] | 1,300,000 | 975 | 470 | 150 | 75 | 1,200,000 | 650 | 950 | 1,200,000 |
| 16F-V5[10] | SP/SP | 1600 | 1600 | 560[11,12] | 560[12] | 200 | 1,400,000 | 1600 | 560 | 175 | 90 | 1,400,000 | 1050 | 1550 | 1,400,000 |
| 20F-V1 | SP/SP | 2000 | 1000 | 650 | 560[12] | 200 | 1,500,000 | 1450 | 560 | 175 | 90 | 1,400,000 | 1000 | 1450 | 1,400,000 |
| 20F-V2 | SP/SP | 2000 | 1000 | 650 | 560[12] | 200 | 1,600,000 | 1450 | 560 | 175 | 90 | 1,400,000 | 1050 | 1550 | 1,400,000 |
| 20F-V3 | SP/SP | 2000 | 1000 | 560[11,12] | 560[12] | 200 | 1,400,000 | 1600 | 560 | 175 | 90 | 1,400,000 | 1000 | 1500 | 1,400,000 |
| 20F-V4[9] | SP/SP | 2000 | 1000 | 650 | 560 | 90[12] | 1,500,000 | 1100 | 470 | 150 | 75 | 1,300,000 | 725 | 950 | 1,300,000 |
| 20F-V5[10] | SP/SP | 2000 | 2000 | 650 | 650 | 200 | 1,600,000 | 1450 | 560 | 175 | 90 | 1,400,000 | 1050 | 1550 | 1,400,000 |
| 22F-V1 | SP/SP | 2200 | 1100 | 650 | 650 | 200 | 1,600,000 | 1600 | 560 | 175 | 90 | 1,500,000 | 1050 | 1650 | 1,500,000 |
| 22F-V2 | SP/SP | 2200 | 1100 | 560[11,12] | 560[12] | 200 | 1,400,000 | 1600 | 560 | 175 | 90 | 1,400,000 | 1000 | 1500 | 1,400,000 |
| 22F-V3 | SP/SP | 2200 | 1100 | 650 | 560[12] | 200 | 1,600,000 | 1500 | 560 | 175 | 90 | 1,400,000 | 1050 | 1500 | 1,400,000 |
| 22F-V4[9] | SP/SP | 2200 | 1100 | 650 | 560[12] | 90[12] | 1,600,000 | 1250 | 470 | 155 | 80 | 1,400,000 | 825 | 1000 | 1,400,000 |
| 22F-V5[10] | SP/SP | 2200 | 2200 | 650 | 650 | 200 | 1,600,000 | 1600 | 560 | 175 | 90 | 1,500,000 | 1050 | 1600 | 1,500,000 |
| 24F-V1 | SP/SP | 2400 | 1200 | 650 | 560[12] | 200 | 1,700,000 | 1500 | 560 | 175 | 90 | 1,500,000 | 1100 | 1350 | 1,500,000 |
| 24F-V2 | SP/SP | 2400 | 1200 | 650 | 650 | 200 | 1,700,000 | 1600 | 560 | 175 | 90 | 1,500,000 | 1100 | 1600 | 1,500,000 |
| 24F-V3 | SP/SP | 2400 | 1200 | 650 | 650 | 200 | 1,800,000 | 1600 | 560 | 175 | 90 | 1,600,000 | 1150 | 1700 | 1,600,000 |
| 24F-V4[9] | SP/SP | 2400 | 1200 | 560 | 560 | 90[12] | 1,700,000 | 1250 | 470 | 155 | 80 | 1,400,000 | 850 | 1050 | 1,400,000 |
| 24F-V5[10] | SP/SP | 2400 | 2400 | 650 | 650 | 200 | 1,700,000 | 1600 | 560 | 175 | 90 | 1,500,000 | 1150 | 1700 | 1,500,000 |
| 24F-V6 | SP/SP | 2400 | 1200 | 650 | 650 | 200 | 1,700,000 | 1500 | 560 | 175 | 90 | 1,500,000 | 1150 | 1750 | 1,500,000 |
| **E-RATED SOUTHERN PINE** | | | | | | | | | | | | | | | |
| 16F-E1 | SP/SP | 1600 | 800 | 560[12] | 560[12] | 200 | 1,600,000 | 1550 | 560 | 175 | 90 | 1,500,000 | 1050 | 1600 | 1,500,000 |
| 16F-E2[9] | SP/SP | 1600 | 800 | 560[12] | 560[12] | 90[12] | 1,600,000 | 950 | 470 | 145 | 75 | 1,300,000 | 700 | 1050 | 1,300,000 |
| 16F-E3[10] | SP/SP | 1600 | 1600 | 560[12] | 560[12] | 200 | 1,600,000 | 1700 | 560 | 175 | 90 | 1,500,000 | 1100 | 1650 | 1,500,000 |
| 20F-E1 | SP/SP | 2000 | 1000 | 560[12] | 560[12] | 200 | 1,700,000 | 1600 | 560 | 175 | 90 | 1,500,000 | 1050 | 1600 | 1,500,000 |
| 20F-E2[9] | SP/SP | 2000 | 1000 | 650 | 650 | 90[12] | 1,600,000 | 1100 | 470 | 150 | 75 | 1,400,000 | 750 | 1000 | 1,400,000 |
| 20F-E3[10] | SP/SP | 2000 | 2000 | 560[12] | 560[12] | 200 | 1,700,000 | 1800 | 560 | 175 | 90 | 1,500,000 | 1150 | 1700 | 1,500,000 |
| 22F-E1 | SP/SP | 2200 | 1100 | 650 | 560[12] | 200 | 1,700,000 | 1600 | 560 | 175 | 90 | 1,500,000 | 1050 | 1650 | 1,500,000 |
| 22F-E2[9] | SP/SP | 2200 | 1100 | 650 | 560[12] | 90[12] | 1,700,000 | 1250 | 470 | 155 | 80 | 1,400,000 | 850 | 1050 | 1,400,000 |
| 22F-E3[10] | SP/SP | 2200 | 2200 | 650 | 650 | 200 | 1,700,000 | 1750 | 560 | 175 | 90 | 1,500,000 | 1150 | 1650 | 1,500,000 |
| 24F-E1 | SP/SP | 2400 | 1200 | 650 | 650 | 200 | 1,800,000 | 1600 | 560 | 175 | 90 | 1,600,000 | 1100 | 1750 | 1,600,000 |
| 24F-E2 | SP/SP | 2400 | 1200 | 650 | 650 | 200 | 1,900,000 | 1700 | 560 | 175 | 90 | 1,600,000 | 1150 | 1700 | 1,600,000 |
| 24F-E3[9] | SP/SP | 2400 | 1200 | 650 | 650 | 90[12] | 1,800,000 | 1300 | 470 | 155 | 80 | 1,500,000 | 950 | 1100 | 1,500,000 |
| 24F-E4[10] | SP/SP | 2400 | 2400 | 650 | 650 | 200 | 1,800,000 | 2000 | 560 | 175 | 90 | 1,600,000 | 1250 | 1750 | 1,600,000 |
| Wet-use factors | | 0.8 | 0.8 | 0.53 | 0.53 | 0.875 | 0.833 | 0.8 | 0.53 | 0.875 | 0.875 | 0.833 | 0.8 | 0.73 | 0.833 |

By permission of NFPA

# Timber Design

## Appendix 2 (cont'd)

**Footnotes**
Applicable to STRUCTURAL GLUED LAMINATED SOFTWOOD TIMBERS:
MEMBERS STRESSED PRIMARILY IN BENDING

1. Design values in this table are based on combinations conforming to AITC 117-84—"Design Standard Specifications for Structural Glued Laminated Timber for Softwood Species", by American Institute of Timber Construction, and manufactured in accordance with American National Standard ANSI/AITC A190.1–1983, "Structural Glued Laminated Timber".
2. The combinations in this table are intended primarily for members stressed in bending due to loads applied perpendicular to the wide faces of the laminations. Design values are tabulated, however, for loading both perpendicular and parallel to the wide faces of the laminations, and for axial loading. For combinations applicable to members loaded primarily axially or parallel to the wide faces of the laminations, see Table 5B.
3. Design values in this table are applicable to members having 4 or more laminations. For members having 2 or 3 laminations, see Table 5B.
4. When moisture content in service will be 16 percent or more, tabulated design values shall be multiplied by the modification factor for wet service conditions, as given in the bottom line of this table.
5. The tabulated design values in bending are applicable to members 12" or less in depth. For members greater than 12" in depth, the requirements of Section 5.3.4 of the National Design Specification apply.
6. The 22F and 24F combinations for members 15" and less in depth may not be readily available and the designer should check on availability prior to specifying. The 16F and 20F combinations are generally available for members 15" and less in depth.
7. The symbols used for species are DF = Douglas Fir-Larch, DFS = Douglas Fir South, HF = Hem-Fir, WW = Western Woods or Canadian softwood species, and SP = Southern Pine (N3 refers to No. 3 structural joists and planks or structural light framing grade). For design values for California Redwood, see AITC 117-84–DESIGN.
8. Design values in this column are for extreme fiber stress in bending when the member is loaded such that the compression zone laminations are subjected to tensile stresses. For more information, see AITC 117-84–DESIGN. The values in this column may be increased 200 psi where end joint spacing restrictions are applied to the compression zone when stressed in tension.
9. These combinations are intended for straight or slightly cambered members for dry use and industrial appearance grade, because they may contain wane. If wane is omitted these restrictions do not apply.
10. These combinations are balanced and are intended for members continuous or cantilevered over supports and provide equal capacity in both positive and negative bending.
11. For bending members greater than 15" in depth, these design values for compression perpendicular to grain are 650 psi on the tension face.
12. These design values may be increased in accordance with AITC 117-84–DESIGN when member conforms with special construction requirements therein. For more information see AITC 117-84–DESIGN.
13. For these combinations manufacturers may substitute E-rated Douglas Fir-Larch laminations that are 200,000 psi higher in modulus of elasticity than the specified E-rated Hem-Fir, with no change in design values.
14. For fastener design, the appropriate timber connector load group, lag bolt and driven fastener load group, and bolt design value can be classified by the design value for compression perpendicular to grain, as shown in the following table:

| | Species classification for fastener design | | |
|---|---|---|---|
| Compression Perpendicular to Grain Design Value $F_{c\perp}$ psi | Timber Connector Load Grouping (NDS Table 8.1A.) | Lag Screw and Driven Fastener Load Grouping (NDS Table 8.1A.) | Bolt Design Values in NDS Table 8.5A |
| 650* | A | II | Column 1 |
| 590 or 560 | B | II | Column 3 |
| 500** | C | III | Column 8 |
| 470 or 375 | C | III | Column 8 |
| 315 | C | III | Column 8 |
| 255 | D | IV | Column 12 |

*For $F_{c\perp}$ = 650 psi for Douglas Fir South, use timber connector Group B, driven fastener group II and bolt values from column 3.
**For $F_{c\perp}$ = 500 psi for Douglas Fir South, use bolt values from column 6.

15. Footnote 5 to Table 5B also applies.
16. The values for horizontal shear, $F_{vyy}$, apply to members manufactured using multiple piece laminations with unbonded edge joints. For members manufactured using single piece laminations or using multiple piece laminations with bonded edge joints the horizontal shear values in the previous column apply.
17. Where Douglas Fir South is used in place of all Western Wood laminations required, the design value for horizontal shear is the same as for combinations using all Douglas Fir-Larch.

By permission of NFPA

## Appendix 3

## TYPICAL DIMENSIONS OF STANDARD LAG SCREWS FOR WOOD

[All dimensions in inches]

- D = Nominal diameter.
- Ds = D = Diameter of shank.
- Dr = Diameter at root of thread.
- W = Width of head across flats.
- H = Height of head.
- L = Nominal length.
- S = Length of shank.
- T = Length of thread.
- E = Length of tapered tip.
- N = Number of threads per inch.

| Nominal length L in inches* | Item | Dimensions of lag screw with nominal diameter D of— | | | | | | | | | | | |
|---|---|---|---|---|---|---|---|---|---|---|---|---|---|
| | | 3/16 | 1/4 | 5/16 | 3/8 | 7/16 | 1/2 | 9/16 | 5/8 | 3/4 | 7/8 | 1 | 1-1/8 | 1-1/4 |
| All lengths | Ds=D | 0.190 | 0.250 | 0.3125 | 0.375 | 0.4375 | 0.500 | 0.5625 | 0.625 | 0.750 | 0.875 | 1.000 | 1.125 | 1.250 |
| | Dr | 0.120 | 0.173 | 0.227 | 0.265 | 0.328 | 0.371 | 0.435 | 0.471 | 0.579 | 0.683 | 0.780 | 0.887 | 1.012 |
| | E | 5/32 | 3/16 | 1/4 | 1/4 | 9/32 | 5/16 | 3/8 | 3/8 | 7/16 | 1/2 | 9/16 | 5/8 | 3/4 |
| | H | 9/64 | 11/64 | 13/64 | 1/4 | 19/64 | 21/64 | 3/8 | 27/64 | 1/2 | 19/32 | 21/32 | 3/4 | 27/32 |
| | W | 9/32 | 3/8 | 1/2 | 9/16 | 5/8 | 3/4 | 7/8 | 15/16 | 1-1/8 | 1-5/16 | 1-1/2 | 1-11/16 | 1-7/8 |
| | N | 11 | 10 | 9 | 7 | 7 | 6 | 6 | 5 | 4-1/2 | 4 | 3-1/2 | 3-1/4 | 3-1/4 |
| 1 | S | 1/4 | 1/4 | 1/4 | 1/4 | 1/4 | 1/4 | | | | | | | |
| | T | 3/4 | 3/4 | 3/4 | 3/4 | 3/4 | 3/4 | | | | | | | |
| | T-E | 19/32 | 9/16 | 1/2 | 1/2 | 15/32 | 7/16 | | | | | | | |
| 1½ | S | 3/8 | 3/8 | 3/8 | 3/8 | 3/8 | 3/8 | | | | | | | |
| | T | 1-1/8 | 1-1/8 | 1-1/8 | 1-1/8 | 1-1/8 | 1-1/8 | | | | | | | |
| | T-E | 31/32 | 15/16 | 7/8 | 7/8 | 27/32 | 13/16 | | | | | | | |
| 2 | S | 1/2 | 1/2 | 1/2 | 1/2 | 1/2 | 1/2 | 1/2 | 1/2 | | | | | |
| | T | 1-1/2 | 1-1/2 | 1-1/2 | 1-1/2 | 1-1/2 | 1-1/2 | 1-1/2 | 1-1/2 | | | | | |
| | T-E | 1-11/32 | 1-5/16 | 1-1/4 | 1-1/4 | 17/32 | 1-3/16 | 1-1/8 | 1-1/8 | | | | | |
| 2½ | S | 1 | 1 | 7/8 | 7/8 | 3/4 | 3/4 | 3/4 | 3/4 | | | | | |
| | T | 1-1/2 | 1-1/2 | 1-5/8 | 1-5/8 | 1-3/4 | 1-3/4 | 1-3/4 | 1-3/4 | | | | | |
| | T-E | 1-11/16 | 1-5/16 | 1-3/8 | 1-3/8 | 1-15/32 | 1-7/16 | 1-3/8 | 1-3/8 | | | | | |
| 3 | S | 1 | 1 | 1 | 1 | 1 | 1 | 1 | 1 | 1 | 1 | 1 | | |
| | T | 2 | 2 | 2 | 2 | 2 | 2 | 2 | 2 | 2 | 2 | 2 | | |
| | T-E | 1-27/32 | 1-13/16 | 1-3/4 | 1-3/4 | 1-23/32 | 1-11/16 | 1-5/8 | 1-5/8 | 1-9/16 | 1-1/2 | 1-7/16 | | |
| 4 | S | 1-1/2 | 1-1/2 | 1-1/2 | 1-1/2 | 1-1/2 | 1-1/2 | 1-1/2 | 1-1/2 | 1-1/2 | 1-1/2 | 1-1/2 | 1-1/2 | 1-1/2 |
| | T | 2-1/2 | 2-1/2 | 2-1/2 | 2-1/2 | 2-1/2 | 2-1/2 | 2-1/2 | 2-1/2 | 2-1/2 | 2-1/2 | 2-1/2 | 2-1/2 | 2-1/2 |
| | T-E | 2-11/32 | 2-5/16 | 2-1/4 | 2-1/4 | 2-7/32 | 2-3/16 | 2-1/8 | 2-1/8 | 2-1/16 | 2 | 1-15/16 | 1-7/8 | 1-3/4 |
| 5 | S | 2 | 2 | 2 | 2 | 2 | 2 | 2 | 2 | 2 | 2 | 2 | 2 | 2 |
| | T | 3 | 3 | 3 | 3 | 3 | 3 | 3 | 3 | 3 | 3 | 3 | 3 | 3 |
| | T-E | 2-27/32 | 2-13/16 | 2-3/4 | 2-3/4 | 2-23/32 | 2-11/16 | 2-5/8 | 2-5/8 | 2-9/16 | 2-1/2 | 2-7/16 | 2-3/8 | 2-1/4 |
| 6 | S | 2-1/2 | 2-1/2 | 2-1/2 | 2-1/2 | 2-1/2 | 2-1/2 | 2-1/2 | 2-1/2 | 2-1/2 | 2-1/2 | 2-1/2 | 2-1/2 | 2-1/2 |
| | T | 3-1/2 | 3-1/2 | 3-1/2 | 3-1/2 | 3-1/2 | 3-1/2 | 3-1/2 | 3-1/2 | 3-1/2 | 3-1/2 | 3-1/2 | 3-1/2 | 3-1/2 |
| | T-E | 3-11/32 | 3-5/16 | 3-1/4 | 3-1/4 | 3-7/32 | 3-3/16 | 3-1/8 | 3-1/8 | 3-1/16 | 3 | 2-15/16 | 2-7/8 | 2-3/4 |
| 7 | S | 3 | 3 | 3 | 3 | 3 | 3 | 3 | 3 | 3 | 3 | 3 | 3 | 3 |
| | T | 4 | 4 | 4 | 4 | 4 | 4 | 4 | 4 | 4 | 4 | 4 | 4 | 4 |
| | T-E | 3-27/32 | 3-13/16 | 3-3/4 | 3-3/4 | 3-23/32 | 3-11/16 | 3-5/8 | 3-5/8 | 3-9/16 | 3-1/2 | 3-7/16 | 3-3/8 | 3-1/4 |
| 8 | S | 3-1/2 | 3-1/2 | 3-1/2 | 3-1/2 | 3-1/2 | 3-1/2 | 3-1/2 | 3-1/2 | 3-1/2 | 3-1/2 | 3-1/2 | 3-1/2 | 3-1/2 |
| | T | 4-1/2 | 4-1/2 | 4-1/2 | 4-1/2 | 4-1/2 | 4-1/2 | 4-1/2 | 4-1/2 | 4-1/2 | 4-1/2 | 4-1/2 | 4-1/2 | 4-1/2 |
| | T-E | 4-11/16 | 4-5/16 | 4-1/4 | 4-1/4 | 4-7/32 | 4-3/16 | 4-1/8 | 4-1/8 | 4-1/16 | 4 | 3-15/16 | 4-1/2 | 3-3/4 |
| 9 | S | 4 | 4 | 4 | 4 | 4 | 4 | 4 | 4 | 4 | 4 | 4 | 4 | 4 |
| | T | 5 | 5 | 5 | 5 | 5 | 5 | 5 | 5 | 5 | 5 | 5 | 5 | 5 |
| | T-E | 4-27/32 | 4-13/16 | 4-3/4 | 4-3/4 | 4-23/32 | 4-11/16 | 4-5/8 | 4-5/8 | 4-9/16 | 4-1/2 | 4-7/16 | 4-3/8 | 4-1/4 |
| 10 | S | 4-3/4 | 4-3/4 | 4-3/4 | 4-3/4 | 4-3/4 | 4-3/4 | 4-3/4 | 4-3/4 | 4-3/4 | 4-3/4 | 4-3/4 | 4-3/4 | 4-3/4 |
| | T | 5-1/4 | 5-1/4 | 5-1/4 | 5-1/4 | 5-1/4 | 5-1/4 | 5-1/4 | 5-1/4 | 5-1/4 | 5-1/4 | 5-1/4 | 5-1/4 | 5-1/4 |
| | T-E | 5-3/32 | 5-1/16 | 5 | 5 | 4-31/32 | 4-15/16 | 4-7/8 | 4-7/8 | 4-13/16 | 4-3/4 | 4-11/16 | 4-5/8 | 4-1/2 |
| 11 | S | 5-1/2 | 5-1/2 | 5-1/2 | 5-1/2 | 5-1/2 | 5-1/2 | 5-1/2 | 5-1/2 | 5-1/2 | 5-1/2 | 5-1/2 | 5-1/2 | 5-1/2 |
| | T | 5-1/2 | 5-1/2 | 5-1/2 | 5-1/2 | 5-1/2 | 5-1/2 | 5-1/2 | 5-1/2 | 5-1/2 | 5-1/2 | 5-1/2 | 5-1/2 | 5-1/2 |
| | T-E | 5-11/32 | 5-9/32 | 5-1/4 | 5-1/4 | 5-7/32 | 5-3/16 | 5-1/8 | 5-1/8 | 5-1/16 | 5 | 4-15/16 | 4-7/8 | 4-3/4 |
| 12 | S | 6 | 6 | 6 | 6 | 6 | 6 | 6 | 6 | 6 | 6 | 6 | 6 | 6 |
| | T | 6 | 6 | 6 | 6 | 6 | 6 | 6 | 6 | 6 | 6 | 6 | 6 | 6 |
| | T-E | 5-27/32 | 5-13/16 | 5-3/4 | 5-3/4 | 5-23/32 | 5-11/16 | 5-5/8 | 5-5/8 | 5-9/16 | 5-1/2 | 5-7/16 | 5-3/8 | 5-1/4 |

*Length of thread T on intervening bolt lengths is the same as that of the next shorter length listed. The length of thread T on standard lag screw lengths L in excess of 12 inches is equal to 1/2 the lag screw length, L/2.

By permission of NFPA

# TIMBER DESIGN

## Appendix 4

## TYPICAL DIMENSIONS FOR THE TIMBER CONNECTORS COVERED IN PART VIII
*Courtesy of TECO Products and Testing Corporation*

### SPLIT RINGS
Dimensions in inches

|  | 2-1/2" | 4" |  | 2-1/2" | 4" |
|---|---|---|---|---|---|
| Split ring: |  |  | Washers, standard: |  |  |
|   Inside diameter at center when closed | 2.500 | 4.000 |   Round, cast or malleable iron, diameter | 2-5/8 | 3 |
|   Thickness of metal at center | 0.163 | 0.193 |   Round, wrought iron (minimum): |  |  |
|   Depth of metal (width of ring) | 0.750 | 1.000 |     Diameter | 1-3/8 | 2 |
|  |  |  |     Thickness | 3/32 | 5/32 |
| Groove: |  |  |   Square plate: |  |  |
|   Inside diameter | 2.56 | 4.08 |     Length of side | 2 | 3 |
|   Width | 0.18 | 0.21 |     Thickness | 1/8 | 3/16 |
|   Depth | 0.375 | 0.50 |  |  |  |
| Bolt hole: |  |  | Projected area: |  |  |
|   Diameter | 9/16 | 13/16 |   Portion of one ring within member, sq. in. | 1.10 | 2.24 |

### SHEAR PLATES
Dimensions in inches

|  | 2-5/8" | 2-5/8" | 4" | 4" |  | 2-5/8" | 2-5/8" | 4" | 4" |
|---|---|---|---|---|---|---|---|---|---|
| Shear plate: |  |  |  |  | Steel strap or shapes for use with shear plates: |  |  |  |  |
|   Material | Pressed steel | Light gage | Malleable iron | Malleable iron |   Steel straps or shapes, for use with shear plates, shall be designed in accordance with accepted engineering practices. |  |  |  |  |
|   Diameter of plate | 2.62 | 2.62 | 4.03 | 4.03 |  |  |  |  |  |
|   Diameter of bolt hole | 0.81 | 0.81 | 0.81 | 0.94 |  |  |  |  |  |
|   Thickness of plate | 0.172 | 0.12 | 0.20 | 0.20 | Hole diameter in straps or |  |  |  |  |
|   Depth of plate | 0.42 | 0.35 | 0.64 | 0.64 |   shapes for bolts | 13/16 | 13/16 | 13/16 | 15/16 |
| Bolt hole—diameter in timber | 13/16 | 13/16 | 13/16 | 15/16 |  |  |  |  |  |
| Washers, standard: |  |  |  |  |  |  |  |  |  |
|   Round, cast or malleable iron, diameter | 3 | 3 | 3 | 3-1/2 |  |  |  |  |  |
|   Round, wrought iron, minimum: |  |  |  |  |  |  |  |  |  |
|     Diameter | 2 | 2 | 2 | 2-1/4 |  |  |  |  |  |
|     Thickness | 5/32 | 5/32 | 5/32 | 11/64 |  |  |  |  |  |
|   Square plate: |  |  |  |  |  |  |  |  |  |
|     Length of side | 3 | 3 | 3 | 3 |  |  |  |  |  |
|     Thickness | 1/4 | 1/4 | 1/4 | 1/4 |  |  |  |  |  |
| Projected area: |  |  |  |  |  |  |  |  |  |
|   Portion of one shear plate within member, sq. in. | 1.18 | 1.00 | 2.58 | 2.58 |  |  |  |  |  |

By permission of NFPA

## Appendix 5
### PROPERTIES OF STRUCTURAL LUMBER

| Nominal size b(inches)d | Standard dressed size (S4S) b(inches)d | Area of Section A | Moment of inertia I | Section modulus S | Weight* in pounds per linear foot of piece when weight of wood per cubic foot equals: | | | | | |
|---|---|---|---|---|---|---|---|---|---|---|
| | | | | | 25 lb. | 30 lb. | 35 lb. | 40 lb. | 45 lb. | 50 lb. |
| 1 x 3 | 3/4 x 2-1/2 | 1.875 | 0.977 | 0.781 | 0.326 | 0.391 | 0.456 | 0.521 | 0.586 | 0.651 |
| 1 x 4 | 3/4 x 3-1/2 | 2.625 | 2.680 | 1.531 | 0.456 | 0.547 | 0.638 | 0.729 | 0.820 | 0.911 |
| 1 x 6 | 3/4 x 5-1/2 | 4.125 | 10.398 | 3.781 | 0.716 | 0.859 | 1.003 | 1.146 | 1.289 | 1.432 |
| 1 x 8 | 3/4 x 7-1/4 | 5.438 | 23.817 | 6.570 | 0.944 | 1.133 | 1.322 | 1.510 | 1.699 | 1.888 |
| 1 x 10 | 3/4 x 9-1/4 | 6.938 | 49.466 | 10.695 | 1.204 | 1.445 | 1.686 | 1.927 | 2.168 | 2.409 |
| 1 x 12 | 3/4 x 11-1/4 | 8.438 | 88.989 | 15.820 | 1.465 | 1.758 | 2.051 | 2.344 | 2.637 | 2.930 |
| 2 x 3 | 1-1/2 x 2-1/2 | 3.750 | 1.953 | 1.563 | 0.651 | 0.781 | 0.911 | 1.042 | 1.172 | 1.302 |
| 2 x 4 | 1-1/2 x 3-1/2 | 5.250 | 5.359 | 3.063 | 0.911 | 1.094 | 1.276 | 1.458 | 1.641 | 1.823 |
| 2 x 5 | 1-1/2 x 4-1/2 | 6.750 | 11.391 | 5.063 | 1.172 | 1.406 | 1.641 | 1.875 | 2.109 | 2.344 |
| 2 x 6 | 1-1/2 x 5-1/2 | 8.250 | 20.797 | 7.563 | 1.432 | 1.719 | 2.005 | 2.292 | 2.578 | 2.865 |
| 2 x 8 | 1-1/2 x 7-1/4 | 10.875 | 47.635 | 13.141 | 1.888 | 2.266 | 2.643 | 3.021 | 3.398 | 3.776 |
| 2 x 10 | 1-1/2 x 9-1/4 | 13.875 | 98.932 | 21.391 | 2.409 | 2.891 | 3.372 | 3.854 | 4.336 | 4.818 |
| 2 x 12 | 1-1/2 x 11-1/4 | 16.875 | 177.979 | 31.641 | 2.930 | 3.516 | 4.102 | 4.688 | 5.273 | 5.859 |
| 2 x 14 | 1-1/2 x 13-1/4 | 19.875 | 290.775 | 43.891 | 3.451 | 4.141 | 4.831 | 5.521 | 6.211 | 6.901 |
| 3 x 1 | 2-1/2 x 3/4 | 1.875 | 0.088 | 0.234 | 0.326 | 0.391 | 0.456 | 0.521 | 0.586 | 0.651 |
| 3 x 2 | 2-1/2 x 1-1/2 | 3.750 | 0.703 | 0.938 | 0.651 | 0.781 | 0.911 | 1.042 | 1.172 | 1.302 |
| 3 x 4 | 2-1/2 x 3-1/2 | 8.750 | 8.932 | 5.104 | 1.519 | 1.823 | 2.127 | 2.431 | 2.734 | 3.038 |
| 3 x 5 | 2-1/2 x 4-1/2 | 11.250 | 18.984 | 8.438 | 1.953 | 2.344 | 2.734 | 3.125 | 3.516 | 3.906 |
| 3 x 6 | 2-1/2 x 5-1/2 | 13.750 | 34.661 | 12.604 | 2.387 | 2.865 | 3.342 | 3.819 | 4.297 | 4.774 |
| 3 x 8 | 2-1/2 x 7-1/4 | 18.125 | 79.391 | 21.901 | 3.147 | 3.776 | 4.405 | 5.035 | 5.664 | 6.293 |
| 3 x 10 | 2-1/2 x 9-1/4 | 23.125 | 164.886 | 35.651 | 4.015 | 4.818 | 5.621 | 6.424 | 7.227 | 8.030 |
| 3 x 12 | 2-1/2 x 11-1/4 | 28.125 | 296.631 | 52.734 | 4.883 | 5.859 | 6.836 | 7.813 | 8.789 | 9.766 |
| 3 x 14 | 2-1/2 x 13-1/4 | 33.125 | 484.625 | 73.151 | 5.751 | 6.901 | 8.051 | 9.201 | 10.352 | 11.502 |
| 3 x 16 | 2-1/2 x 15-1/4 | 38.125 | 738.870 | 96.901 | 6.619 | 7.943 | 9.266 | 10.590 | 11.914 | 13.238 |
| 4 x 1 | 3-1/2 x 3/4 | 2.625 | 0.123 | 0.328 | 0.456 | 0.547 | 0.638 | 0.729 | 0.820 | 0.911 |
| 4 x 2 | 3-1/2 x 1-1/2 | 5.250 | 0.984 | 1.313 | 0.911 | 1.094 | 1.276 | 1.458 | 1.641 | 1.823 |
| 4 x 3 | 3-1/2 x 2-1/2 | 8.750 | 4.557 | 3.646 | 1.519 | 1.823 | 2.127 | 2.431 | 2.734 | 3.038 |
| 4 x 4 | 3-1/2 x 3-1/2 | 12.250 | 12.505 | 7.146 | 2.127 | 2.552 | 2.977 | 3.403 | 3.828 | 4.253 |
| 4 x 5 | 3-1/2 x 4-1/2 | 15.750 | 26.578 | 11.813 | 2.734 | 3.281 | 3.828 | 4.375 | 4.922 | 5.469 |
| 4 x 6 | 3-1/2 x 5-1/2 | 19.250 | 48.526 | 17.646 | 3.342 | 4.010 | 4.679 | 5.347 | 6.016 | 6.684 |
| 4 x 8 | 3-1/2 x 7-1/4 | 25.375 | 111.148 | 30.661 | 4.405 | 5.286 | 6.168 | 7.049 | 7.930 | 8.811 |
| 4 x 10 | 3-1/2 x 9-1/4 | 32.375 | 230.840 | 49.911 | 5.621 | 6.745 | 7.869 | 8.933 | 10.117 | 11.241 |
| 4 x 12 | 3-1/2 x 11-1/4 | 39.375 | 415.283 | 73.828 | 6.836 | 8.203 | 9.570 | 10.938 | 12.305 | 13.672 |
| 4 x 14 | 3-1/2 x 13-1/4 | 46.375 | 678.475 | 102.411 | 8.047 | 9.657 | 11.266 | 12.877 | 14.485 | 16.094 |
| 4 x 16 | 3-1/2 x 15-1/4 | 53.375 | 1034.418 | 135.66 | 9.267 | 11.121 | 12.975 | 14.828 | 16.682 | 18.536 |
| 5 x 2 | 4-1/2 x 1-1/2 | 6.750 | 1.266 | 1.688 | 1.172 | 1.406 | 1.641 | 1.875 | 2.109 | 2.344 |
| 5 x 3 | 4-1/2 x 2-1/2 | 11.250 | 5.859 | 4.688 | 1.953 | 2.344 | 2.734 | 3.125 | 3.516 | 3.906 |
| 5 x 4 | 4-1/2 x 3-1/2 | 15.750 | 16.078 | 9.188 | 2.734 | 3.281 | 3.828 | 4.375 | 4.922 | 5.469 |
| 5 x 5 | 4-1/2 x 4-1/2 | 20.250 | 34.172 | 15.188 | 3.516 | 4.219 | 4.922 | 5.675 | 6.328 | 7.031 |
| 6 x 1 | 5-1/2 x 3/4 | 4.125 | 0.193 | 0.516 | 0.716 | 0.859 | 1.003 | 1.146 | 1.289 | 1.432 |
| 6 x 2 | 5-1/2 x 1-1/2 | 8.250 | 1.547 | 2.063 | 1.432 | 1.719 | 2.005 | 2.292 | 2.578 | 2.865 |
| 6 x 3 | 5-1/2 x 2-1/2 | 13.750 | 7.161 | 5.729 | 2.387 | 2.865 | 3.342 | 3.819 | 4.297 | 4.774 |
| 6 x 4 | 5-1/2 x 3-1/2 | 19.250 | 19.651 | 11.229 | 3.342 | 4.010 | 4.679 | 5.347 | 6.016 | 6.684 |
| 6 x 6 | 5-1/2 x 5-1/2 | 30.250 | 76.255 | 27.729 | 5.252 | 6.302 | 7.352 | 8.403 | 9.453 | 10.503 |
| 6 x 8 | 5-1/2 x 7-1/2 | 41.250 | 193.359 | 51.563 | 7.161 | 8.594 | 10.026 | 11.458 | 12.891 | 14.323 |
| 6 x 10 | 5-1/2 x 9-1/2 | 52.250 | 392.963 | 82.729 | 9.071 | 10.885 | 12.700 | 14.514 | 16.328 | 18.142 |
| 6 x 12 | 5-1/2 x 11-1/2 | 63.250 | 697.068 | 121.229 | 10.981 | 13.177 | 15.373 | 17.569 | 19.766 | 21.962 |
| 6 x 14 | 5-1/2 x 13-1/2 | 74.250 | 1127.672 | 167.063 | 12.891 | 15.469 | 18.047 | 20.625 | 23.203 | 25.781 |
| 6 x 16 | 5-1/2 x 15-1/2 | 85.250 | 1706.776 | 220.229 | 14.800 | 17.760 | 20.720 | 23.681 | 26.641 | 29.601 |
| 6 x 18 | 5-1/2 x 17-1/2 | 96.250 | 2456.380 | 280.729 | 16.710 | 20.052 | 23.394 | 26.736 | 30.078 | 33.420 |
| 6 x 20 | 5-1/2 x 19-1/2 | 107.250 | 3398.484 | 348.563 | 18.620 | 22.344 | 26.068 | 29.792 | 33.516 | 37.240 |
| 6 x 22 | 5-1/2 x 21-1/2 | 118.250 | 4555.086 | 423.729 | 20.530 | 24.635 | 28.741 | 32.847 | 36.953 | 41.059 |
| 6 x 24 | 5-1/2 x 23-1/2 | 129.250 | 5948.191 | 506.229 | 22.439 | 26.927 | 31.415 | 35.903 | 40.391 | 44.878 |
| 8 x 1 | 7-1/4 x 3/4 | 5.438 | 0.255 | 0.680 | 0.944 | 1.133 | 1.322 | 1.510 | 1.699 | 1.888 |
| 8 x 2 | 7-1/4 x 1-1/2 | 10.875 | 2.039 | 2.719 | 1.888 | 2.266 | 2.643 | 3.021 | 3.398 | 3.776 |
| 8 x 3 | 7-1/4 x 2-1/2 | 18.125 | 9.440 | 7.552 | 3.147 | 3.776 | 4.405 | 5.035 | 5.664 | 6.293 |
| 8 x 4 | 7-1/4 x 3-1/2 | 25.375 | 25.904 | 14.803 | 4.405 | 5.286 | 6.168 | 7.049 | 7.930 | 8.811 |
| 8 x 6 | 7-1/2 x 5-1/2 | 41.250 | 103.984 | 37.813 | 7.161 | 8.594 | 10.026 | 11.458 | 12.891 | 14.323 |
| 8 x 8 | 7-1/2 x 7-1/2 | 56.250 | 263.672 | 70.313 | 9.766 | 11.719 | 13.672 | 15.625 | 17.578 | 19.531 |
| 8 x 10 | 7-1/2 x 9-1/2 | 71.250 | 535.859 | 112.813 | 12.370 | 14.844 | 17.318 | 19.792 | 22.266 | 24.740 |
| 8 x 12 | 7-1/2 x 11-1/2 | 86.250 | 950.547 | 165.313 | 14.974 | 17.969 | 20.964 | 23.958 | 26.953 | 29.948 |
| 8 x 14 | 7-1/2 x 13-1/2 | 101.250 | 1537.734 | 227.813 | 17.578 | 21.094 | 24.609 | 28.125 | 31.641 | 35.156 |
| 8 x 16 | 7-1/2 x 15-1/2 | 116.250 | 2327.422 | 300.313 | 20.182 | 24.219 | 28.255 | 32.292 | 36.328 | 40.365 |
| 8 x 18 | 7-1/2 x 17-1/2 | 131.250 | 3349.609 | 382.813 | 22.786 | 27.344 | 31.901 | 36.458 | 41.016 | 45.573 |
| 8 x 20 | 7-1/2 x 19-1/2 | 146.250 | 4634.297 | 475.313 | 25.391 | 30.469 | 35.547 | 40.625 | 45.703 | 50.781 |
| 8 x 22 | 7-1/2 x 21-1/2 | 161.250 | 6211.484 | 577.813 | 27.995 | 33.594 | 39.193 | 44.792 | 50.391 | 55.990 |
| 8 x 24 | 7-1/2 x 23-1/2 | 176.250 | 8111.172 | 690.313 | 30.599 | 36.719 | 42.839 | 48.958 | 55.078 | 61.198 |

*Weight in lb/Ft$^3$ = 62.4 x Specific gravity (See Table 8.1A)

By permission of NFPA

# Appendix 5 (cont'd)

## PROPERTIES OF STRUCTURAL LUMBER

| Nominal size $b$(inches)$d$ | Standard dressed size (S4S) $b$(inches)$d$ | Area of Section $A$ | Moment of inertia $I$ | Section modulus $S$ | Weight* in pounds per linear foot of peice when weight of wood per cubic foot equals: | | | | | |
|---|---|---|---|---|---|---|---|---|---|---|
| | | | | | 25 lb. | 30 lb. | 35 lb. | 40 lb. | 45 lb. | 50 lb. |
| 10 x 1  | 9-1/4 x 3/4     | 6.938   | 0.325     | 0.867    | 1.204  | 1.445  | 1.686  | 1.927  | 2.168  | 2.409  |
| 10 x 2  | 9-1/4 x 1-1/2   | 13.875  | 2.602     | 3.469    | 2.409  | 2.891  | 3.372  | 3.854  | 4.336  | 4.818  |
| 10 x 3  | 9-1/4 x 2-1/2   | 23.125  | 12.044    | 9.635    | 4.015  | 4.818  | 5.621  | 6.424  | 7.227  | 8.030  |
| 10 x 4  | 9-1/4 x 3-1/2   | 32.375  | 33.049    | 18.885   | 5.621  | 6.745  | 7.869  | 8.993  | 10.117 | 11.241 |
| 10 x 6  | 9-1/2 x 5-1/2   | 52.250  | 131.714   | 47.896   | 9.071  | 10.885 | 12.700 | 14.514 | 16.328 | 18.142 |
| 10 x 8  | 9-1/2 x 7-1/2   | 71.250  | 333.984   | 89.063   | 12.370 | 14.844 | 17.318 | 19.792 | 22.266 | 24.740 |
| 10 x 10 | 9-1/2 x 9-1/2   | 90.250  | 678.755   | 142.896  | 15.668 | 18.802 | 21.936 | 25.069 | 28.203 | 31.337 |
| 10 x 12 | 9-1/2 x 11-1/2  | 109.250 | 1204.026  | 209.396  | 18.967 | 22.760 | 26.554 | 30.347 | 34.141 | 37.934 |
| 10 x 14 | 9-1/2 x 13-1/2  | 128.250 | 1947.797  | 288.563  | 22.266 | 26.719 | 31.172 | 35.625 | 40.078 | 44.531 |
| 10 x 16 | 9-1/2 x 15-1/2  | 147.250 | 2948.068  | 380.396  | 25.564 | 30.677 | 35.790 | 40.903 | 46.016 | 51.128 |
| 10 x 18 | 9-1/2 x 17-1/2  | 166.250 | 4242.836  | 484.896  | 28.863 | 34.635 | 40.408 | 46.181 | 51.953 | 57.726 |
| 10 x 20 | 9-1/2 x 19-1/2  | 185.250 | 5870.109  | 602.063  | 32.161 | 38.594 | 45.026 | 51.458 | 57.891 | 64.323 |
| 10 x 22 | 9-1/2 x 21-1/2  | 204.250 | 7867.879  | 731.896  | 35.460 | 42.552 | 49.644 | 56.736 | 63.828 | 70.920 |
| 10 x 24 | 9-1/2 x 23-1/2  | 223.250 | 10274.148 | 874.396  | 38.759 | 46.510 | 54.262 | 62.014 | 69.766 | 77.517 |
| 12 x 1  | 11-1/4 x 3/4    | 8.438   | 0.396     | 1.055    | 1.465  | 1.758  | 2.051  | 2.344  | 2.637  | 2.930  |
| 12 x 2  | 11-1/4 x 1-1/2  | 16.875  | 3.164     | 4.219    | 2.930  | 3.516  | 4.102  | 4.688  | 5.273  | 5.859  |
| 12 x 3  | 11-1/4 x 2-1/2  | 28.125  | 14.648    | 11.719   | 4.883  | 5.859  | 6.836  | 7.813  | 8.789  | 9.766  |
| 12 x 4  | 11-1/4 x 3-1/2  | 39.375  | 40.195    | 22.969   | 6.836  | 8.203  | 9.570  | 10.938 | 12.305 | 13.672 |
| 12 x 6  | 11-1/2 x 5-1/2  | 63.250  | 159.443   | 57.979   | 10.981 | 13.177 | 15.373 | 17.569 | 19.766 | 21.962 |
| 12 x 8  | 11-1/2 x 7-1/2  | 86.250  | 404.297   | 107.813  | 14.974 | 17.969 | 20.964 | 23.958 | 26.953 | 29.948 |
| 12 x 10 | 11-1/2 x 9-1/2  | 109.250 | 821.651   | 172.979  | 18.967 | 22.760 | 26.554 | 30.347 | 34.141 | 37.934 |
| 12 x 12 | 11-1/2 x 11-1/2 | 132.250 | 1457.505  | 253.479  | 22.960 | 27.552 | 32.144 | 36.736 | 41.328 | 45.920 |
| 12 x 14 | 11-1/2 x 13-1/2 | 155.250 | 2357.859  | 349.313  | 26.953 | 32.344 | 37.734 | 43.125 | 48.516 | 53.906 |
| 12 x 16 | 11-1/2 x 15-1/2 | 178.250 | 3568.713  | 460.479  | 30.946 | 37.135 | 43.325 | 49.514 | 55.703 | 61.892 |
| 12 x 18 | 11-1/2 x 17-1/2 | 201.250 | 5136.066  | 586.979  | 34.939 | 41.927 | 48.915 | 55.903 | 62.891 | 69.878 |
| 12 x 20 | 11-1/2 x 19-1/2 | 224.250 | 7105.922  | 728.813  | 38.932 | 46.719 | 54.505 | 62.292 | 70.078 | 77.865 |
| 12 x 22 | 11-1/2 x 21-1/2 | 247.250 | 9524.273  | 885.979  | 42.925 | 51.510 | 60.095 | 68.681 | 77.266 | 85.851 |
| 12 x 24 | 11-1/2 x 23-1/2 | 270.250 | 12437.129 | 1058.479 | 46.918 | 56.302 | 65.686 | 75.069 | 84.453 | 93.837 |
| 14 x 2  | 13-1/4 x 1-1/2  | 19.875  | 3.727     | 4.969    | 3.451  | 4.141  | 4.831  | 5.521  | 6.211  | 6.901  |
| 14 x 3  | 13-1/4 x 2-1/2  | 33.125  | 17.253    | 13.802   | 5.751  | 6.901  | 8.051  | 9.201  | 10.352 | 11.502 |
| 14 x 4  | 13-1/4 x 3-1/2  | 46.375  | 47.34     | 27.052   | 8.047  | 9.657  | 11.266 | 12.877 | 14.485 | 16.094 |
| 14 x 6  | 13-1/2 x 5-1/2  | 74.250  | 187.172   | 68.063   | 12.891 | 15.469 | 18.047 | 20.625 | 23.203 | 25.781 |
| 14 x 8  | 13-1/2 x 7-1/2  | 101.250 | 474.609   | 126.563  | 17.578 | 21.094 | 24.609 | 28.125 | 31.641 | 35.156 |
| 14 x 10 | 13-1/2 x 9-1/2  | 128.250 | 964.547   | 203.063  | 22.266 | 26.719 | 31.172 | 35.625 | 40.078 | 44.531 |
| 14 x 12 | 13-1/2 x 11-1/2 | 155.250 | 1710.984  | 297.563  | 26.953 | 32.344 | 37.734 | 43.125 | 48.516 | 53.906 |
| 14 x 14 | 13-1/2 x 13-1/2 | 182.250 | 2767.922  | 410.063  | 31.641 | 37.969 | 44.297 | 50.625 | 56.953 | 63.281 |
| 14 x 16 | 13-1/2 x 15-1/2 | 209.250 | 4189.359  | 540.563  | 36.328 | 43.594 | 50.859 | 58.125 | 65.391 | 72.656 |
| 14 x 18 | 13-1/2 x 17-1/2 | 236.250 | 6029.297  | 689.063  | 41.016 | 49.219 | 57.422 | 65.625 | 73.828 | 82.031 |
| 14 x 20 | 13-1/2 x 19-1/2 | 263.250 | 8341.734  | 855.563  | 45.703 | 54.844 | 63.984 | 73.125 | 82.266 | 91.406 |
| 14 x 22 | 13-1/2 x 21-1/2 | 290.250 | 11180.672 | 1040.063 | 50.391 | 60.469 | 70.547 | 80.625 | 90.703 | 100.781 |
| 14 x 24 | 13-1/2 x 23-1/2 | 317.250 | 14600.109 | 1242.563 | 55.078 | 66.094 | 77.109 | 88.125 | 99.141 | 110.156 |
| 16 x 3  | 15-1/4 x 2-1/2  | 38.125  | 19.857    | 15.885   | 6.619  | 7.944  | 9.267  | 10.592 | 11.915 | 13.240 |
| 16 x 4  | 15-1/4 x 3-1/2  | 53.375  | 54.487    | 31.135   | 9.267  | 11.121 | 12.975 | 14.828 | 16.682 | 18.536 |
| 16 x 6  | 15-1/2 x 5-1/2  | 85.250  | 214.901   | 78.146   | 14.800 | 17.760 | 20.720 | 23.681 | 26.641 | 29.601 |
| 16 x 8  | 15-1/2 x 7-1/2  | 116.250 | 544.922   | 145.313  | 20.182 | 24.219 | 28.255 | 32.292 | 36.328 | 40.365 |
| 16 x 10 | 15-1/2 x 9-1/2  | 147.250 | 1107.443  | 233.146  | 25.564 | 30.677 | 35.790 | 40.903 | 46.016 | 51.128 |
| 16 x 12 | 15-1/2 x 11-1/2 | 178.250 | 1964.463  | 341.646  | 30.946 | 37.135 | 43.325 | 49.514 | 55.703 | 61.892 |
| 16 x 14 | 15-1/2 x 13-1/2 | 209.250 | 3177.984  | 470.813  | 36.328 | 43.594 | 50.859 | 58.125 | 65.391 | 72.656 |
| 16 x 16 | 15-1/2 x 15-1/2 | 240.250 | 4810.004  | 620.646  | 41.710 | 50.052 | 58.394 | 66.736 | 75.078 | 83.420 |
| 16 x 18 | 15-1/2 x 17-1/2 | 271.250 | 6922.523  | 791.146  | 47.092 | 56.510 | 65.929 | 75.347 | 84.766 | 94.184 |
| 16 x 20 | 15-1/2 x 19-1/2 | 302.250 | 9577.547  | 982.313  | 52.474 | 62.969 | 73.464 | 83.958 | 94.453 | 104.948 |
| 16 x 22 | 15-1/2 x 21-1/2 | 333.250 | 12837.066 | 1194.146 | 57.856 | 69.427 | 80.998 | 92.569 | 104.141 | 115.712 |
| 16 x 24 | 15-1/2 x 23-1/2 | 364.250 | 16763.086 | 1426.646 | 63.238 | 75.885 | 88.533 | 101.181 | 113.828 | 126.476 |
| 18 x 6  | 17-1/2 x 5-1/2  | 96.250  | 242.630   | 88.229   | 16.710 | 20.052 | 23.394 | 26.736 | 30.078 | 33.420 |
| 18 x 8  | 17-1/2 x 7-1/2  | 131.250 | 615.234   | 164.063  | 22.786 | 27.344 | 31.901 | 36.458 | 41.016 | 45.573 |
| 18 x 10 | 17-1/2 x 9-1/2  | 166.250 | 1250.338  | 263.229  | 28.863 | 34.635 | 40.408 | 46.181 | 51.953 | 57.726 |
| 18 x 12 | 17-1/2 x 11-1/2 | 201.250 | 2217.943  | 385.729  | 34.939 | 41.927 | 48.915 | 55.903 | 62.891 | 69.878 |
| 18 x 14 | 17-1/2 x 13-1/2 | 236.250 | 3588.047  | 531.563  | 41.016 | 49.219 | 57.422 | 65.625 | 73.828 | 82.031 |
| 18 x 16 | 17-1/2 x 15-1/2 | 271.250 | 5430.648  | 700.729  | 47.092 | 56.510 | 65.929 | 75.347 | 84.766 | 94.184 |
| 18 x 18 | 17-1/2 x 17-1/2 | 306.250 | 7815.754  | 893.229  | 53.168 | 63.802 | 74.436 | 85.069 | 95.703 | 106.337 |
| 18 x 20 | 17-1/2 x 19-1/2 | 341.250 | 10813.359 | 1109.063 | 59.245 | 71.094 | 82.943 | 94.792 | 106.641 | 118.490 |
| 18 x 22 | 17-1/2 x 21-1/2 | 376.250 | 14493.461 | 1348.229 | 65.321 | 78.385 | 91.450 | 104.514 | 117.578 | 130.642 |
| 18 x 24 | 17-1/2 x 23-1/2 | 411.250 | 18926.066 | 1610.729 | 71.398 | 85.677 | 99.957 | 114.236 | 128.516 | 142.795 |

*Weight in lb/Ft$^3$ = 62.4 x Specific gravity (See Table 8.1A)

By permission of NFPA

## Appendix 6

## Curved Glulam Factors

| $\beta$ | A | B | C |
|---|---|---|---|
| (0.0) | (0.0) | (0.2500) | (0.0) |
| 2.5 | 0.0079 | 0.1747 | 0.1284 |
| 5.0 | 0.0174 | 0.1251 | 0.1939 |
| 7.5 | 0.0279 | 0.0937 | 0.2162 |
| 10.0 | 0.0391 | 0.0754 | 0.2119 |
| 15.0 | 0.0629 | 0.0619 | 0.1722 |
| 20.0 | 0.0893 | 0.0608 | 0.1393 |
| 25.0 | 0.1214 | 0.0605 | 0.1238 |
| 30.0 | 0.1649 | 0.0603 | 0.1115 |

## 13. REFERENCES

American Institute of Timber Construction, <u>Glulam Systems</u>, AITC, 333 West Hampden Avenue, Englewood, CO 80110, 1980

American Plywood Association, <u>Plywood Design Specification</u>, APA, P.O. Box 11700, Tacoma, Washington, 1980

American Plywood Association, <u>Design and Fabrication of Plywood-Lumber Beams -- Supplement 2</u>, APA, P.O. Box 11700, Tacoma, Washington, 1982

American Plywood Association, <u>Plywood Diaphragm Construction</u>, APA, P.O. Box 11700, Tacoma, Washington, 1970

American Society for Testing and Materials, <u>Annual Book of ASTM Standards, Part 22, Wood; Adhesives</u>, ASTM, 1916 Race Street, Philadelphia, PA, 19103

Bodig, Jozsef, and Jayne, Benjamin A., <u>Mechanics of Wood and Wood Composites</u>, Van Nostrand Reinhold Company, New York City, New York, 1982

Breyer, Donald E., <u>Design of Wood Structures</u>, McGraw-Hill, New York City, New York, 1980

Diekmann, Edward F., <u>Design of Diaphragms</u>, chapter in Volume 4 of EMMSE Series, Materials Research Laboratory, The Pennsylvania State University, University Park, PA, 16802, 1984

Gurfinkel, German, <u>Wood Engineering</u>, 2nd edition, Kendall/Hunt Publishing Co., Dubuque, Iowa, 1981

Hoyle, Robert J., Jr., <u>Wood Technology in the Design of Structures</u>, 4th Edition, Mountain Press Publishing Company, Missoula, MT

International Conference of Building Officials, <u>Uniform Building Code</u>, 1982 Edition, ICBO, 5360 South Workman Mill Road, Whittier, CA 90601

National Forest Products Association, <u>National Design Specification for Wood Construction</u>, 1982 Edition, NFPA, 1619 Massachusetts Avenue, N.W., Washington, DC 20036

Timber Engineering Company, <u>Design Manual for TECO Timber Connector Construction</u>, TECO, 5530 Wisconsin Avenue, Chevy Chase, MD, 20815

Forest Products Laboratory, United States Department of Agriculture, <u>Wood Handbook (Wood as an Engineering Material)</u>, Forest Service, Handbook No. 72, Revised 1974

# INDEX

## A

Allowable bending stresses 105, 107
Allowable lateral loads 30
Allowable stress tables 19
Allowable stresses 21
Allowable stresses in plywood 126
American Plywood Association 119
Amount of moisture 8
Angle of load to grain 39
Axial deflection 89
Axial loadings 86
Axial members 94

## B

Beam deflections 112
Beam lengths 106
Beam slenderness categories 107
Beams 20
Bearing 22
Bearing stresses at loads and supports 112
Bending deflection increase to account for shear 144
Bending members 105
Bending stress 90, 105
Bolt diameter 63
Bolt hole 63
Bolts 36, 45
Bound water 8
Buckling 87
Built-up beams 139
Built-up columns 104

## C

Capacity reduction with multiple fasteners 36
Cellulose 4
Checked beams 111
Checks 6
Combinations of load types 25
Combined loadings 92
Compression members 95
Compression parallel to the grain 22
Compression perpendicular to the grain 22
Compression wood 7
Condition of use 38
Condition of use factors 15
Condition of use factors for connections 28
Conductivity 11
Coniferous 3
Connection capacity 28
Connection details 50
Connections 23
Connector spacing 65, 66
Connectors 23, 61
Contraction 11
Correction factors 14
Cross grain 5, 13
Curvature factor 109
Curved bending members 108

## D

Dead load 16
Deciduous 3
Defects 4
Deflection 91, 112, 143
Density 7
Diaphragms 131, 132
Dimension lumber 20
Distortion 10
Double-shear assumption 46

## E

Eccentric loading 94
Edge distance 65
Edges 20
Effective beam length 106
Effective column length 89
End distance 65
Equilibrium moisture content 9
Euler equation 87
Euler's critical stress 87
Expansion 11
Extractives 8
Extreme fiber stress in bending 21

## F

Faces 20
Fiber saturation point 9
Fire 11
Five percent exclusion limit 14
Flanges 142
Form factors 107
Free water 8
Full-size members 20
Full-size testing 17

## G

Glue-laminated allowable stress 22
Glue-laminated beams 108
Glue-laminated members 21
Grading 14, 18
Green 8

## H

Hankinson formula 25
Hardwood 3
Heartwood 8
Holding members 23
Horizontal shear 21

## I

Impact load 16
Infiltrates 8
Insurance rates 11
Intermediate beams 107
Intermediate columns 98

## K

Knots 4

## L

Lag bolt 40
Lag screw spacing 41
Lag screws 36, 40
Lateral buckling 105
Lateral load 24, 131
Lateral loads in lag screws 40
Lateral stability 92
Length-width ratio 137
Lignin 4
Limit-state design 13
Load combinations 16
Load-deformation curve 29
Load duration 15, 26
Load duration factors 15
Load tables 46
Loads near supports 110
Long beams 107
Long columns 98
Longitudinal axis 12
Lumber flanges 142

## M

Machine-stress-rated 18
Metal members in connection 47
Metal side plates 30
Modulus of elasticity 22
Moisture content 8, 15
Multiple fastener factor 37

## N

Nail and spike dimensions 28
Nails 26, 29
Net section 39
Notched beams 110

## O

Orthotropic material 12
Oval knot 5
Oven-dry 9

## P

Parallel to grain 25
Perpendicular to grain 25
Pilot hole 40
Plain-sawed 12
Plywood 119
Plywood grades 119
Plywood section modulus 124
Plywood section properties 121
Plywood stamp 120
Plywood webs 143
Poisson's ratio 13
Posts 20

## Q

Quarter-sawed 12

## R

Radial axis 12
Radial compression 109
Radial tension 109
Reaction wood 7
Repetitive use member 21
Rolling shear 125
Round knot 5

## S

Safety factor 14
Sapwood 8
Shakes 6
Shear plates 36, 61
Shear stress 90
Shear stresses in bending members 109
Shear walls 131, 133
Short beams 107
Short columns 97
Shrinkage 9
Side members 23
Sides 20
Single shear 46
Size factor 108
Slenderness ratio 88, 97
Slope of grain 6
Small clear specimen method 13
Snow load 16
Softwood 3
Spaced columns 103
Spacing 66
Spacing requirements 52
Species groupings for connection design 27
Specific gravity 7, 10
Spike knot 5
Spikes 26
Split rings 36, 61
Splits 6
Staggered fasteners 36
Stress-graded 18
Stringers 20
Strong axis 12
Structural plywood 119

## T

Tangential axis 12
TECO 23, 61
Tension members 94
Tension parallel to the grain 21
Tension wood 7
Thermal conductivities 11
Thermal properties 10
Timber Engineering Company 23, 61
Timbers 20
Toe-nailed connections 30
Transverse loading 90

## U

Uniform Building Code 137

## W

Web 143
Wind and seismic load 16
Withdrawal 24
Withdrawal loads 40
Wood density 7
Working stresses 13

**RESERVED FOR FUTURE USE**

# DON'T GAMBLE!
## These books will extract Every Last Point from the examination for you

## Engineering Law, Design Liability, and Professional Ethics

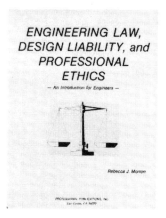

The most difficult problems are essay questions about management, ethics, professional responsibility, and law. Since these questions can ask for definitions of terms you're not likely to know, it is virtually impossible to fake it by rambling on. And yet, these problems are simple if you have the right resources. If you don't feel comfortable with such terms as comparative negligence, discovery proceedings, and strict liability in tort, you should bring **Engineering Law, Ethics, and Liability** with you to the examination.

None of this material is in your review manual. And, nothing from your review manual has been duplicated here.

8½" × 11", soft cover, 120 pages, $15.80 (includes postage).

(ISBN 0-932276-37-7)

## Expanded Interest Tables

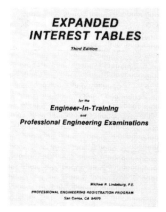

There's nothing worse than knowing how to solve a problem but not having the necessary data. Engineering Economics problems are like that. You might know how to do a problem, but where do you get interest factors for non-integer interest rates? **Expanded Interest Tables** will prove indispensible for such problems. It has pages for interest rates starting at ¼% and going to 25% in ¼% increments. Factors are given for up to 100 years. There's no other book like it. So, if you want to be prepared for an Engineering Economy problem with 11.75% interest, you need **Expanded Interest Tables**.

8½" × 11", soft cover, 106 pages, $15.80, including postage.

(ISBN 0-932276-35-0)

*Phone and mail orders are accepted.*

**PROFESSIONAL PUBLICATIONS, INC.**
1250 Fifth Avenue
Belmont, CA 94002
(415) 593-9119